1/23/84

ORGANIC COATINGS

ORGANIC COATINGS
Science and Technology, Volume 5

edited by

Geoffrey D. Parfitt

Department of Chemical Engineering
Carnegie-Mellon University
Pittsburgh, Pennsylvania

Angelos V. Patsis

Department of Chemistry
Institute in Science and Technology
State University of New York
New Paltz, New York

MARCEL DEKKER, INC.   New York and Basel

Library of Congress Cataloging in Publication Data
main entry under title:

Organic Coatings, science and technology.

   "Based on papers originally presented at the Seventh
International Conference in Organic Coatings Science
and Technology in Athens, Greece"--Pref.
   1. Plastic coatings--Congresses. I. Parfitt, G. D.
II. Patsis, Angelos V. III. International Conference
in Organic Coatings Science and Technology (7th :
1981 : Athens, Greece) IV. Series: Organic coatings ;
v. 5.
TP1175.S607 1983   668.4   83-14338
ISBN 0-8247-1905-0

MARCEL DEKKER, INC.
270 Madison Avenue, New York, New York 10016

Current printing (last digit):
10 9 8 7 6 5 4 3 2 1

PRINTED IN THE UNITED STATES OF AMERICA

75

PREFACE

    This publication is based on papers originally presented at the Seventh International Conference in Organic Coatings Science and Technology in Athens, Greece. With the 1981 Conference, the Athens Conference (to give it its popular name) has firmly established itself as the premier annual scientific and technological conference on coatings in Europe. The conference is modeled on the U. S. Gordon Conference. There were 21 invited and 5 short contributed papers.

    The latest developments in the field of organic coatings were presented with talks in the area of binders, with special emphasis on epoxies, powder and water-soluble coatings, and pigments. The sciences of rheology and adhesion as applied to the coatings technology was also discussed. Novel, but practical coatings application techniques was the subject of some presentations.

    Most of the papers presented at the conference are collected in this volume. The topics covered range from developmental efforts to fundamental research.

    We would like to express our appreciation to all those who helped in preparing this volume, especially the authors.

G. Parfitt
A. V. Patsis

CONTENTS

v

Contents

CONTRIBUTORS

RONALD S. BAUER   Westhollow Research Center, Shell Development
    Center, Houston, Texas

E. G. BELDER   Research and Development, Scado B. V., Zwolle, The
    Netherlands

GUY C. BELL, Jr.   Fabrics and Finishes Department, E. I. du Pont de
    Nemours & Co., Inc., Marshall Laboratory, Philadelphia,
    Pennsylvania

JAMES V. CRIVELLO   General Electric Corporate Research and Develop-
    ment Center, Schenectady, New York

E. M. A. de JONG   Sikkens B. V. — A Company of Akzo Coatings,
    Sassenheim, The Netherlands

V. GARZITTO   Istituto Chimica, Universita degli Studi di Udine, Udine,
    Italy

A. HEERINGA   Automotive Research Department, Sikkens B. V.,
    Sassenheim, The Netherlands

JEROME HOCHBERG   Fabrics and Finishes Department, E. I. du Pont
    de Nemours & Co., Inc., Experimental Station, Wilmington,
    Delaware

KENNETH L. HOY   Union Carbide Corporation, South Charleston, West
    Virginia

KLAUS HUNGER   Dyestuffs, Pigments, and Intermediates Research, Hoechst AG, Frankfurt am Main, West Germany

LARS IGETOFT   Department of Chemistry and Center for Surface and Coatings Research, Lehigh University, Sinclair Laboratory #7, Bethlehem, Pennsylvania

JU KUMANOTANI   The University of Tokyo, Institute of Industrial Science, Tokyo, Japan

J. KUNNEN   Sikkens B. V. — A Company of Akzo Coatings, Sassenheim, The Netherlands

HENRY LEIDHEISER, Jr.   Department of Chemistry and Center for Surface and Coatings Research, Lehigh University, Bethlehem, Pennsylvania

P. J. LLOYD   Depatment of Chemical Engineering, University of Technology, Loughborough, Leicestershire, England

KURT MERKLE   Marketing Pigments, Hoechst AG, Frankfurt am Main, West Germany

SATOSHI OKUDA   Department of Chemical Engineering, Faculty of Engineering, Doshisha University, Kyoto, Japan

A. PAPO   Istituto de Chimica, Università degli Studi di Udine, Udine, Italy

T. SATOH   Research Laboratory, Kansai Paint Co., Ltd., Hiratsuka, Kanagawa, Japan

PALLE SØRENSON   Printing Ink Development, Sadolin & Holmblad Ltd., Glostrup, Denmark

S. A. STACHOWIAK   Surface Coatings Section, Resins Department, Koninklijke/Shell-Laboratorium (Shell Research B. V.), Amsterdam, The Netherlands

F. STURZI   Istituto de Chimica, Università degli Studi di Udine, Udine, Italy

R. van der LINDE   Research and Development, Scado B. V., Zwolle, The Netherlands

P. J. G. van HENSBERGEN   Automotive Research Department, Sikkens B. V., Sassenheim, The Netherlands

BRIAN VINCENT   Department of Physical Chemistry, University of Bristol, Bristol, England

WENDY WANG   Department of Chemistry and Center for Surface and Coatings Research, Lehigh University, Bethlehem, Pennsylvania

KEITH WEBER   Department of Chemistry and Center for Surface and Coatings Research, Lehigh University, Bethlehem, Pennsylvania

JAMES T. K. WOO   Glidden Coatings and Resins, Dwight P. Joyce
     Research Center, Strongsville, Ohio

ROLF ZIMMERMANN   Hoechst AG, Frankfurt am Main, West Germany

ORGANIC COATINGS

RECENT DEVELOPMENTS IN EPOXY RESINS

Ronald S. Bauer

Shell Development Company
Westhollow Research Center
Houston, Texas

INTRODUCTION

Superimposed on the evolutionary advancements that are intrinsic traits of the coatings industry, ecological pressures and the need to conserve energy are creating demands for major technological changes in the coatings industry. For example, the United States Environmental Protection Agency (EPA) has issued guidelines they want enacted by individual regions in the United States before December 1982, limiting the amount of solvent in industrial coatings on an industry-by-industry basis. These guidelines are summarized in Table 1.1, and as can be seen, with only a few exceptions most industries will be required to limit the volatile organic compounds emitted to 340 g/L of paint or less. The 340 g/L limit very roughly corresponds to 60%-65% volume solids, depending on the composition of the particular coating.

The model rule for marine coatings adopted by the California Air Resources Board (CARB) will eventually prohibit coatings from containing more than 295 g of volatile organic material per liter of coating as applied. The CARB model rule dealing with so called architectural coatings, which includes by CARB definition industrial maintenance coatings, proposes a 250 g/L limit. The 295 g/L and 250 g/L limits very roughly correspond to 65% and 70% volume solids, respectively. There is already much activity in at least 25 other states to adopt high solids regulations similar to the CARB model rules.

1

TABLE 1.1
EPA Guidelines
Recommended Limitations[a] for
Various Industrial Coatings

Coating Use	Recommended Limitations Volatile Organic Components	
	g/L	Approx %v
Automobile primers	230	70–75
Automobile topcoats	340	60–65
Large appliances	340	60–65
Can coatings		
Roller coatings	340	60–65
Interior spray	500	45
Side seam spray	660	30
Coil coatings	310	65
Metal furniture	360	60
Magnetic wire insulation	200	75–80

[a]Recommended limitations obtained from a publication of the
United States Environmental Protection Agency. The volume
percentages were estimated from the EPA recommended
limitations.

It is estimated then that in the United States, the market percent for
all types of industrial coatings resins going into solution coatings (coatings
having less than 70% v solids) will drop during the period from 1977 to 1985
from 74% to 33%. The market for resins in all types of high solids industrial coatings (coatings having more 70% v solids) will increase in this
period from 6% to 23% (Table 1.2). These changes, though spawned by
government legislation to reduce solvent-vapor emissions, will certainly
result in other benefits, such as lower energy consumption during curing
and baking operations as well as substantially reduced solvent costs.

These pressures have not gone unnoticed by the suppliers of epoxy
resins, curing agents, and manufacturers of epoxy resin coatings. Over
the past 5 years, epoxy resin sales in the United States have experienced
an annual growth rate of about 8%/year. In 1980, production of epoxy resins
exceeded 288 million pounds and domestic sales were greater than 254
million pounds with the markets being divided almost equally between protective coatings and structural end uses [1](Figure 1.1). Although small
when compared to the volume of alkyds and polyesters used as binders by

TABLE 1.2
Technology Forecast
Industrial Coating Resins

	Percent Market Share		
	1977	1981	1985
Solution (< 70% v)	74	48	33
High solids (> 70% v)	6	16	23
Powder	3	5	6
Water–borne	14	25	31
Other (UV)	3	6	7

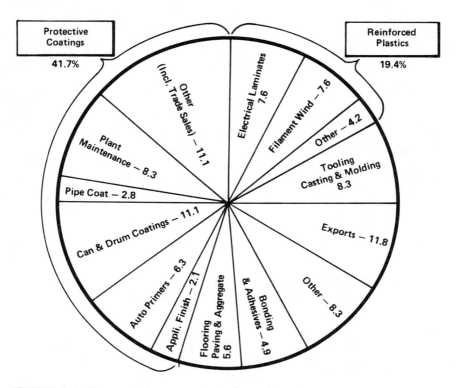

FIGURE 1.1 Epoxy resin consumption by end–use (1980).

the coatings industry in the United States, it still constitutes a significant quantity of premium high performance organic coatings. The following describes some of the recent interesting developments in epoxy resin technology that should assure continued epoxy resin usage in the premium coatings market.

WATER-BORNE EPOXY COATINGS

Water-borne coatings have always offered considerable promise as a means of obtaining ecologically compliant coatings for industrial applications. Unfortunately, they have not entirely lived up to that promise, and water-borne epoxy resin coatings have been no exception. Many attempts have been made to develop such coatings, and some of these have met with a certain amount of success for particular applications, such as electrodeposited primers, industrial maintenance coatings, and coatings for beer and beverage cans. The patents and publications dealing with new water-borne epoxy based coating resins and epoxy coatings are too numerous to deal with here in any depth. Therefore, this discussion will be limited to the approaches to water-borne epoxy finishes that are currently showing the greatest technical and commercial success.

Water-Borne Baking Finishes

The application where epoxy resin based water-borne coatings probably have achieved their greatest success to date is in automotive primers applied by electrodeposition (ED). Electrodeposition of coatings using anodic systems is well established technology, having the proven advantages of efficiency of coating utilization, being essentially nonpolluting, and capable of a high degree of automation. More recently, however, cathodic systems have been replacing the anodic type because of the increased corrosion resistance obtained with cathodic coatings while retaining the advantages of the anodic process. The preferred system appears to be based on:

(1)   a hydroxyl-containing reactive product of a primary or
        secondary amine and a polyepoxide, which is solubilized
        with an acid, and

(2)   a blocked isocyanate crosslinker, which is stable at
        ambient temperatures in the presence of the hydroxyl-
        containing material but reactive with the epoxy adduct at
        elevated temperatures.

Even in the area of cathodic electrodeposition the literature is now extensive. The following illustrates some of the types of systems proposed for solubilizing the hydroxyl containing portion of the system. In the first example, the terminal epoxy groups of a polyglycidyl ether are reacted with a primary or secondary amine [2] to give an adduct that can be subsequently solubilized by addition of acid, such as or acetic or lactic acids.

AMINE FUNCTIONAL POLYMERS

$$
\underset{O}{\overset{O}{\triangle}}\ CH_2-CH-CH_2-O-R-O-CH_2-\overset{O}{\overset{\triangle}{CH-CH_2}} \quad + \quad R_1R_2N-H \quad \longrightarrow
$$

$$
\underset{R_2}{\overset{R_1}{|}}N-CH_2-\underset{OH}{\overset{|}{CH}}-CH_2-O-R-O-CH_2-\underset{OH}{\overset{|}{CH}}-CH_2-\underset{R_2}{\overset{R_1}{|}}N \quad \xrightarrow[\text{Acid Solution}]{\text{Acetic}} \quad \sim CH-CH_2-\underset{R_2}{\overset{R_1}{\overset{|}{N}}}-H^{\oplus} \quad ACO^{\ominus}
$$

$$
R = -\text{[diphenyl structure]}-
$$

R = [benzene ring]-C(CH₃)₂-[benzene ring]-[O-CH₂CH(OH)-CH₂-O-[benzene ring]-C(CH₃)₂-[benzene ring]]ₙ

R_1 and R_2 = Hydrogen and/or Alkyl

In a second approach, the terminal epoxy groups are adducted with a tertiary amine salt or an internal zwitterion formed, for example by the reaction of a secondary amine and a conjugated unsaturated carboxylic acid [3].

QUARTERNARY AMMONIUM SALT PROCESS

$$
\underset{O}{\overset{O}{\triangle}}\ CH_2-CH-CH_2-O-R-O-CH_2-\overset{O}{\overset{\triangle}{CH-CH_2}} \quad + \quad 2\left[R_2'N-H\right]^{\oplus} \quad + \quad 2R''-\underset{O}{\overset{\overset{}{}}{C}}O^{\ominus}
$$

$$
\downarrow H_2O
$$

$$
\left[\overset{R'}{\underset{R'}{\overset{|}{R'-\overset{\oplus}{N}}}}-CH_2-\underset{OH}{\overset{|}{CH}}-CH_2-O-R-O-CH_2-\underset{OH}{\overset{|}{CH}}-CH_2-\underset{R'}{\overset{R'}{\overset{|}{N}}}-R'^{\oplus} \right] \quad + \quad 2R''-\underset{O}{\overset{}{C}}-O^{\ominus}
$$

$$R = \text{—}\underset{CH_3}{\overset{CH_3}{\underset{|}{\overset{|}{C}}}}\text{—}\left[\text{—O-CH}_2\text{-CH-CH}_2\text{-O—}\underset{OH}{}\right]\text{—}\underset{CH_3}{\overset{CH_3}{\underset{|}{\overset{|}{C}}}}\text{—}\right]_n$$

R' and R" = Alkyl Groups

Also terminal epoxy groups of polyepoxides have been reacted with sulfides and phosphines to give products, which in the presence of an acid result in water dispersible quaternary sulfonium [4] and phosphonium salts [5], respectively.

$$\sim\text{CH-CH}_2\text{-}\underset{R_2}{\overset{R_1}{S}}^{\oplus}\ RCO_2^{\ominus} \qquad \sim\text{CH-CH}_2\text{-}\underset{R_3}{\overset{R_1}{P}}\text{-}R_2{}^{\oplus}RCO_2^{\ominus}$$
$$\underset{OH}{} \qquad\qquad\qquad\qquad \underset{OH}{}$$

SULFONIUM SALT   PHOSPHONIUM SALT

A unique feature of many of these cationic electrodeposition systems is that they are cured with blocked isocyanates. The cure of aminoplast resins typically used as crosslinkers in anodic systems is retarded in these cationic systems by the influence of the amine functionality. The blocked isocyanates are reported to be diurethanes from toluene diisocyanate with, for example, 2-ethylhexanol. Cure schedules for these systems range from about 45 min at 350°F to 10 min at 380°F.

Anodic electrodeposition binders have been reported with much improved corrosion resistance over the current state-of-the-art systems [6a]. Development of these improved ED binders was based on the principle that hydroxyl functionality promotes binder adhesion to the substrate, and consequently improves corrosion resistance as illustrated in Table 1.3. The data summarized in the table indicates good salt-spray resistance on bare steel is obtained with coatings containing between 200 and 400 meq hydroxyl groups per 100 g resin, and it appears that salt spray resistance is unrelated to the degree of crosslinking of the systems.

A typical approach to hydroxyl rich binders of this type is to react the terminal epoxy groups of a solid epoxy resin with stiochiometric quantities of hydroxy acids, such as lactic or dimethylol proprionic acid (DMPA).

TABLE 1.3

Properties of Epoxy Resin/Hexamethoxymethylmelamine (HMMM)
Coatings on Bare Steel[a]

Epoxy Resin[b] /HMMM Weight Ratio	Methylethyl Ketone Resistance	Salt-Spray Resistance[c]		Residual OH Meq/100 g
		1-Day	8-Day	
99/1	Poor	10	9	350
95/5	Good	10	9	305
90/10	Good	$9\frac{1}{2}$	7	250
85/15	Very good	2	0	190
67/33	Very good	0	0	$\simeq 0$

[a] Clear 25 μm coatings applied by spraying from organic solvent solutions
and stoved at 180 °C for 30 min. Systems contained 0.5% w of p-toluene
sulfonic acid as catalyst.

[b] Epoxy resin with Epoxide Equivalent Weight (EEW) range of 1650 to 2050.

[c] Ratings: 10 = unaffected, 9 = 5 mm loss of adhesion from scratch,
8 = 10 mm loss of adhesion, 7 = 15 mm loss of adhesion, etc., 0 = total
loss of adhesion.

PREPARATION OH-RICH EPOXY RESIN
BINDER FOR E/D COATINGS

EPOXY   OH-RICH
RESIN   BACKBONE

The resulting hydroxyl rich backbones are then further reacted with a cyclic anhydride to introduce carboxyl functionality that can subsequently be used as sites for solubilization. Binders with optimum flow, bath stability, and outstanding salt spray resistance are obtained if equimolar amounts of trimellitic anhydride and a glycidyl ester of an α-branched saturated fatty acid are incorporated into the system.

It was found that a polyester type adduct of 1 mole trimellitic an-hydride (TMA) and 1.04 mole of glycidyl ester reacts with hydroxyl-rich resins by transesterification; the polyester is cleaved and bound to the hydroxyl-rich backbone, thereby introducing -COOH groups [6b].

REACTION OF A POLYESTER TYPE ADDUCT OF TMA/GLYCIDYL ESTER WITH AN OH-RICH BACKBONE

$$\boxed{\textbf{OH-Rich Backbone}}\text{-O-CO}\underset{\overset{\displaystyle}{}}{\bigcirc}\overset{\displaystyle \text{COOH}}{\underset{\displaystyle \text{COO-CH}_2\text{-CH-CH}_2\text{-OOC-C}_9\text{H}_{19}}{}}$$

$$\text{COO-CH}_2\text{-}\overset{\displaystyle \overset{\text{OH}}{|}}{\text{CH}}\text{-CH}_2\text{-OOC-C}_9\text{H}_{19}$$

When crosslinked with melamine or phenolic resins, binders of this type show a good balance of ED behavior, mechanical properties and bath stability. Salt-spray resistance of these films is comparable to a commercial low-molecular weight polybutadiene (LMPB) system, which have superior salt-spray resistance on non-phosphated steel.

Although not under the extreme environmental pressures as automotive primers, considerable development activity is evidenced in the area of water-borne epoxy coatings for metal cans (Table 1.1). Interest in this area is understandable when one considers that about 50 billion beer and beverage cans are manufactured annually in the United States. A very high percentage of these cans are coated with epoxy-resin based systems. Epoxy coatings have gained broad acceptance over the years in this application because of their outstanding adhesion to a broad spectrum of substrates, a low organoleptic effect on beer, and a high resistance to low pH carbonated beverages.

A standard technique for obtaining water-dispersable polymers is through incorporation of pendent carboxyl groups attached to a resin molecule, such as epoxy fatty-acid ester through addition of an anhydride, which results in an ester linkage. These ester linkages as well as those formed by the fatty acid are quite vulnerable to hydrolytic attack, due to the basicity of the system imparted by the solubilizing amine. Thus, on storage, the carboxyl groups are split from the resin via ester cleavage, resulting in a gradual decrease in miscibility and ultimately leading to phase separation.

Epoxy Ester/ Anhydride Adduct

➤ Points of Hydrolytic Instability

FA= Fatty Acid Chain

Anhydride Adduct

Water soluble epoxy binders with improved hydrolytic stability have been obtained by reacting low molecular weight epoxy resins with para-amino-benzoic acid (PABA)[7]. Although both amines and carboxylic acids react readily with epoxy functionality, at low temperatures the amine functionality of the PABA will react while leaving the carboxyl essentially unreacted.

EPOXY RESIN/PABA REACTION

The amine solubilized epoxy resin/PABA binders can readily be cured with bake schedules of 200°C for 2 min using a melamine-formaldehyde (CYMEL 370) crosslinking resin. The formulation and properties of a typical clear container coating are given in Table 1.4. The hydrolytic stability of these systems on storage was found to be outstanding and is illustrated in Table 1.5 with data obtained on a one-year-old sample [8].

An adaption of the epoxy resin/PABA system using a slightly higher molecular weight epoxy resin has been formulated into a water-borne automotive primer. The salt-spray resistance of this primer was found to be superior to that of a state-of-the-art commercial solvent-borne epoxy based automotive primer.

TABLE 1.4
Water-Borne Epoxy/PABA Baking System
Cured with CYMEL 370

Dispersion Properties	
Non-volatile, % w	26.3
Visc. Gardner Holdt	"F"
Water/organic by volume	87/13
pH	8.0-8.5
PABA resin/CYMEL 370 (solids basis)	85/15
Film Properties, Baked 2 min at 400°F	
Beer pasteurization	Pass
MEK double rubs	>100
Flex	1T

TABLE 1.5
Hydrolytic Stability of the Water-Borne
PABA/Epoxy Baking System

Epoxy/PABA System	Fresh		Aged 1 Year	
Solids content, measured, % w (after 30 min/160°C)	24.7		24.6	
Viscosity at 23°C, Ford cup #4, sec	67		81	
Substrate	Al.	Tinpl.	Al.	Tinpl.
Flexibility, wedge bend, mm failure	28	31	28	34
Solvent resistance, MEK double rubs	←			→

Another interesting approach to hydrolytically stable water-borne
epoxy vehicles has been reported that involves graft polymerization of a
styrene-methacrylic acid copolymer onto the aliphatic backbone of the
epoxy resin [9]. These epoxy-acrylic copolymers can then be neutralized
with bases, such as dimenthylethanolamine, to give products which easily

disperse in water to give stable coating systems. Grafting is accomplished using benzoyl-peroxide as an initiator and is presumed to occur as follows:

WATER-BORNE EPOXY GRAFT POLYMERS

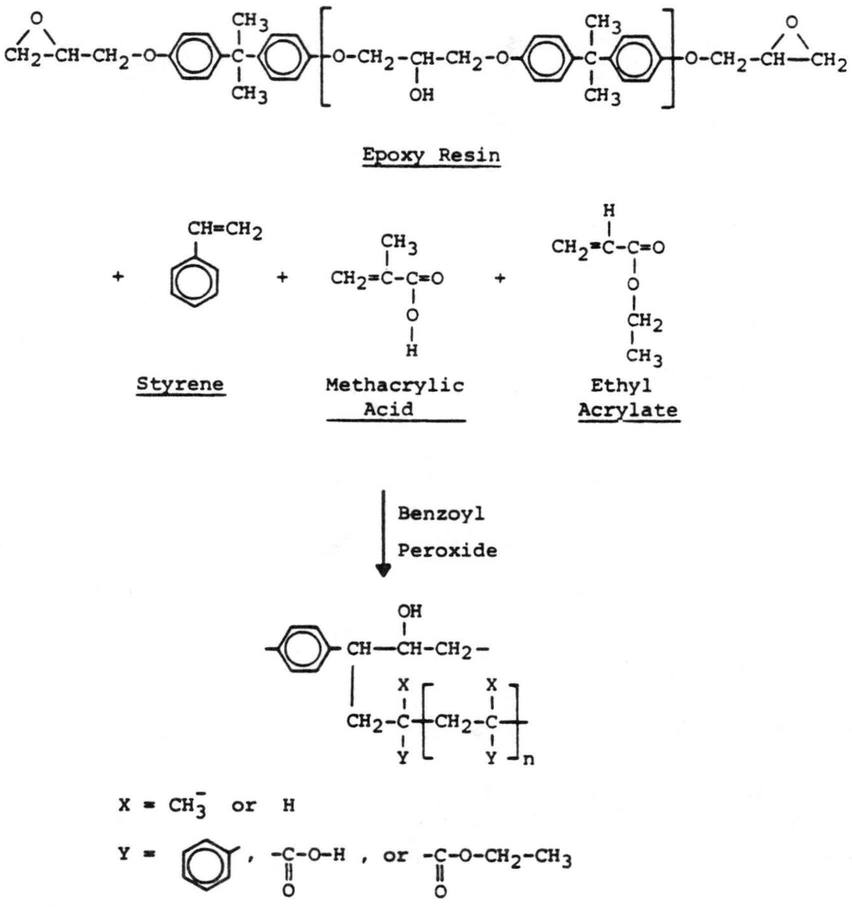

Epoxy Resin

Styrene   Methacrylic   Ethyl
                   Acid   Acrylate

Benzoyl
Peroxide

$X = CH_3^-$ or H

$Y = $, $-C-O-H$, or $-C-O-CH_2-CH_3$

These copolymers are believed to be mixtures of (1) graft copolymer of the unsaturated monomers onto the epoxy backbone, (2) ungrafted epoxy resin and (3) ungrafted addition polymers of the unsaturated monomers. The dispersed resin then can be cured with an aminoplast resin (CYMEL 370) to give coatings with properties that make them suitable for carbonated beverage and beer containers. The hydrolytic stability of these dispersions is illustrated in Table 1.6.

TABLE 1.6
Hydrolytic Stability Water-Borne
Epoxy Graft Polymers

System	Initial Viscosity (sec)[a]	Initial pH	% Neutralization	Viscosity (sec)[a] after 8 mo. at 49°C	pH after 8 mo. at 49°C
A	26	8.35	85	22	8.28
B	33	8.43	85	27	8.4
C	21	8.55	90	17	8.45

a
#4 Ford cup

Water-Borne Ambient-Cure Finishes

A new family of ambient-cure epoxy coatings, which are based on a water-reducible amine functional acrylic copolymer are a novel departure from classical polyamide and polyamine curing systems. An early system in this family consists of an acidified addition copolymer containing carboxyl groups and pendent aminoester groups formed by aminoethylating the carboxyl groups with an excess alkylenimine having the formula [10]

$$-\overset{\overset{\displaystyle O}{\|}}{C}-O-\left[CH-CH-NH_2\right]$$
$$\left[\underset{R_1 \ \ R_2}{}\right]$$

The backbone of the curing agent is formed through free radical polymerization of vinyl monomers, such as acrylic acid, ethyl acrylate, and styrene. These are then amino alkylated with an imine. Upon neutralization the polymer becomes water soluble. An enormous number of possibilities exist. For example, the vinyl acid and its relative amount can be varied, as can the amount of amine and the neutralizing acid, thus allowing the crosslink density of the resulting film to be varied over a wide range.

Also the epoxy component can be varied to achieve a range of film characteristics. Recommended is a liquid diglycidyl ether of bisphenol A (DGEBA) for chemical resistance and a liquid aliphatic diepoxide for flexibility. New epoxy resin technology based on a saturated bisphenol-A now

ACID SOLUBILIZED ACRYLIC/AMINE CURING AGENT

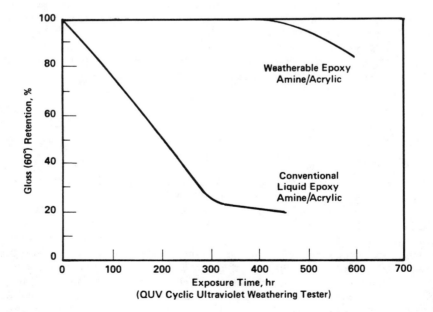

offers the opportunity to obtain water-borne epoxy films with these acrylic
backbone curing agents having excellent gloss retention and non-chalking
characteristics [11].   The much improved gloss retention over the corresponding system formulated with a conventional bisphenol-A epoxy resin is
shown in Figure 1.2.

FIGURE 1.2 Accelerated weathering of water-borne ambient-cure
weatherable epoxy/amine acrylic.

These acidified copolymers are not without their drawbacks, however. As the copolymer product is an acid salt, it is generally not suitable for coating unprimed metal substrates since such acid salts usually cause flash rusting. Another problem is that anionic additives, such as defoamers, pigment dispersants, and colorants cannot be used nor can basic pigments like carbonates.

An amine functionalized, water-reducible acrylic system has been developed which is claimed to minimize the above disadvantages. In this case, the polymeric backbone is also comprised of free radically polymerized vinyl monomers, including an unsaturated carboxylic acid [12]. Only a portion of the carboxylic acid groups are reacted with an alkylenimine to form pendent amine groups and a portion of the unreacted carboxylic acid groups are neutralized to form pendent salt groups. The composition of this base solubilized acrylic/amine epoxy curing agent can be represented schematically as follows:

BASE SOLUBILIZED ACRYLIC/AMINE CURING AGENT

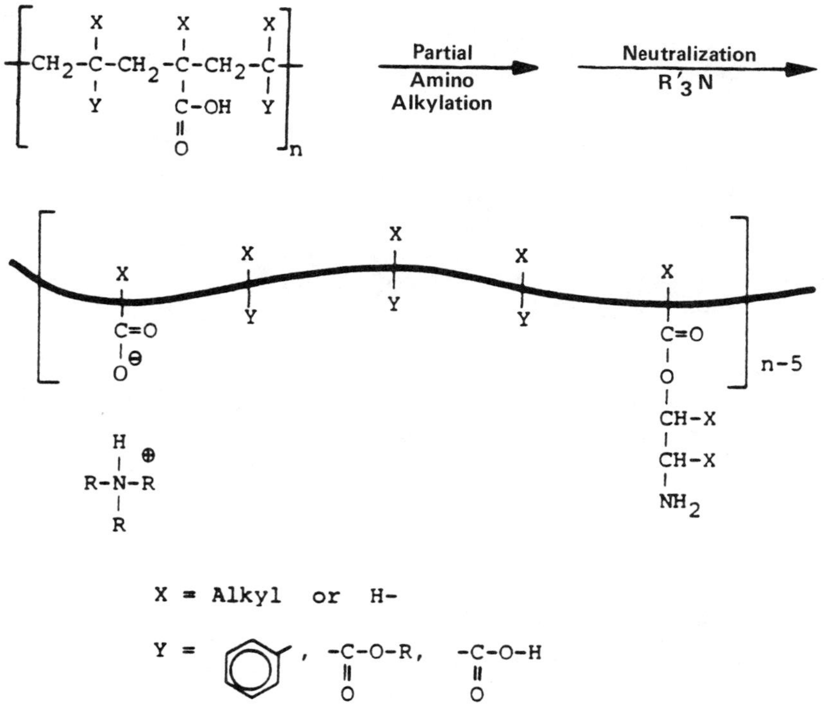

X = Alkyl or H-

Y = , -C-O-R,   -C-O-H
                          ‖   ‖
                          O   O

R = Aliphatic or Cycloaliphatic Radicals

It has been shown, however, that some care must be exercised in the selection of vinyl monomers in order to avoid storage stability problems with this type of acrylic/amine curing agent. For example, a polymer backbone based on monomers, such as ethyl acrylate and methacrylic acid after amination and neutralization showed either a major increase in viscosity or gelled after heat aging for 16 h at 200°F. The poor stability was found to be a result of the reaction between an amine group and an ester forming an amide cross-link. Systems with improved storage stability have now been obtained by incorporation of "aminolysis-resistant" groups, such as isobutylmethacrylate, cyclohexylmethacrylate, or isobornyl methacrylate [13].

Although epoxy/carboxylic acid systems do not generally give acceptable cures at ambient temperatures, an acid functional acrylic copolymer has been developed for ambient-cure, water-borne epoxy coatings [14]. The system is solubilized by addition of tertiary amine, which is claimed to be needed for stability and to catalyze the cure.

BASE SOLUBILIZED ACID/ACRYLIC CURING AGENT

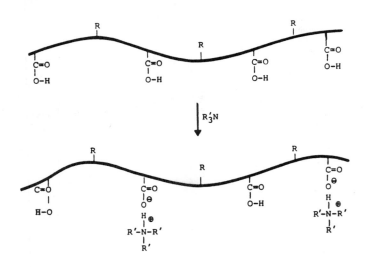

Tertiary Amine Neutralization

· Solubilizes Acrylic Resin

· Catalyzes Cure

· Stabilizes System

TABLE 1.7

Acid Resistance of Water–Borne Acrylic/Amine and
Acid/Acrylic Cured Epoxy Resin Systems
(24 h Spot Test[a])

| | Acid/Acrylic | | Acrylic/Amine | |
	Bisphenol–A Epoxy	Weatherable Epoxy	Bisphenol–A Epoxy	Wheatherable Epoxy
20% HCl	N. A.	N. A.	N. A.	Blisters/6M
5% HCl	N. A.	N. A.	Blisters/8D	Blisters/8D
10% H_2SO_4	N. A.	N. A.	Blisters/4M	Blisters/6MD
10% Acetic acid	Blisters/8MD	Blisters/8M	Blisters/2D	Blisters/2M
40% H_3PO_4	N. A.	N. A.	Softened	Softened

[a]ASTM D714–5 (D714–56, reapproved 1970).
N. A. = Not Affected.
Blisters 10 = None, 0 = very large, VD = Very dense; D = Dense;
M = Medium, F = Few.

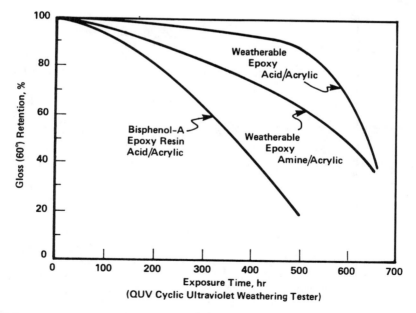

FIGURE 1.3   Accelerated weathering of water-borne, ambient-cure
weatherable epoxy/acrylic amine.

Like the acrylic amine described above, films with properties approaching those of acrylic urethanes, including excellent gloss retention, can be obtained when used in combination with a new experimental weatherable epoxy resin (Figure 1.3) [14]. These reactive carboxylic acid systems also seem to be less sensitive to attack by acid, as evidenced in Table 1.7, than the amine cured systems.

HIGHER SOLIDS EPOXY COATINGS

As it has become more evident that water-borne coatings are not a panacea, more attention seems to be focused on high solids coatings as a means of obtaining compliance systems. As indicated earlier, high solids industrial coatings of all types are projected to increased in volume from 6% to 23% of the market between 1977 and 1985. The shift toward higher solids could accelerate as the cost of solvents and energy continue to increase. For example, as can be seen in Figure 1.4, increasing the volume solids of a coating system from 40% to 70% volume solids reduces the

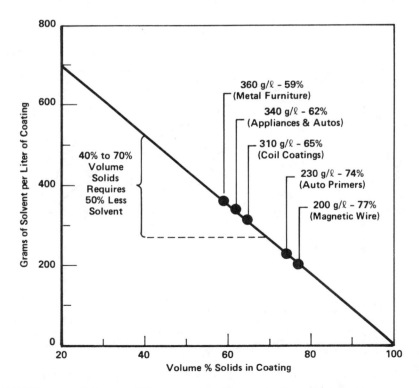

FIGURE 1.4 Volume solids vs. solvent content of organic coatings.

weight of the solvent required by about 50%. Lower solvent emissions from high solids coatings should also reduce fuel costs, since less makeup air for baking ovens is required.

Higher Solids Baking Finishes

Typically epoxy baking systems are based on higher molecular weight epoxy resins cured with an aminoplast or phenoplast resin. These higher molecular weight resins have the following structure:

TYPICAL EPOXY RESIN

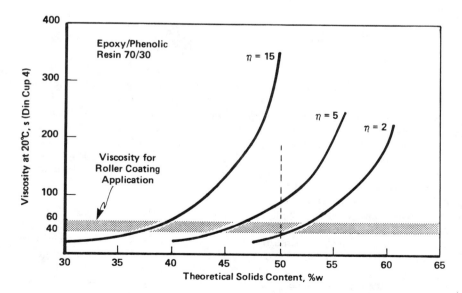

n = 0 to 25

The high molecular weight resins are used primarily because of the high degree of formability (adhesion plus flexibility) they impart to the coating. Figure 1.5 shows the relation between viscosity and solids content of an

FIGURE 1.5   Relationship between viscosity and solids content for selected epoxy resin molecular weights.

epoxy/phenolic system formulated with epoxy resins with varying n-values
[8]. A 50% w solids system, which is applicable by roller coater, would
have to be based on resins with an n between 2 and 5. These systems,
however, would lack the flexibility required for a can lacquer.

A series of flexible resins, based on low molecular epoxy resins
modified with dimer acid, has been developed that offer the potential for
formulating high solids epoxy coatings approaching the performance of the
higher molecular weight systems which are:

MODIFIED EPOXY RESIN FOR HIGH SOLIDS BAKING SYSTEMS

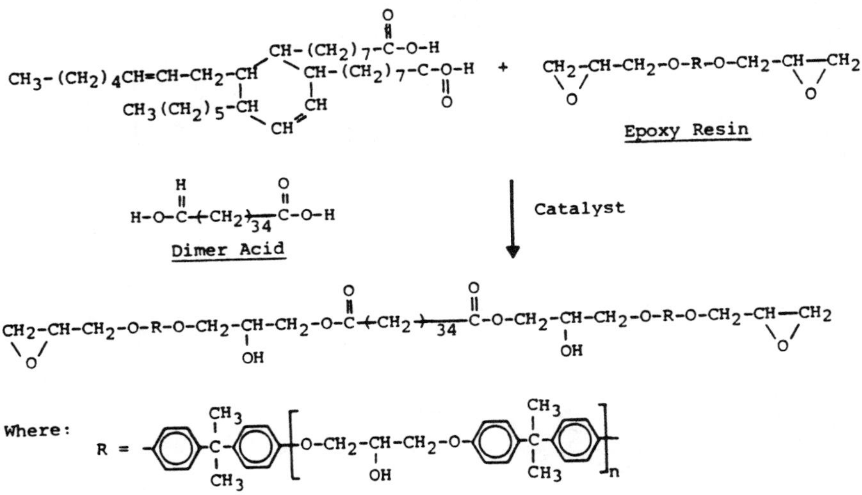

Epoxy Resin

Dimer Acid   Catalyst

Where:

The physical properties of these resins are given in Table 1.8 along
with a typical clear container coating formulation, as well as the performance characteristics of the resulting films. These systems were designed
to be applied by roller coater at temperatures around 50°C; however, the
coatings catalyzed as indicated in the table with levels of phosphoric acid
(0.9% w based on resin solids) sufficient to give adequate cure exhibit poor
storage stability as illustrated in Figure 1.6. Stable systems have been
obtained using p-toluene sulfonic acid and a new FDA acceptable benzene-
sulfonic acid type catalyst.

When catalyzed with higher levels of phosphoric acid (4 phr) these
epoxy/dimer acid resins, surprisingly, with both urea-formaldehyde and
melamine-formaldehyde crosslinking resins give coating systems that have
been found to be stable for up to 45 days at room temperature [15] (Figure
1.7). These systems have been formulated into high solids automotive
primers that give satisfactory cures at temperatures as low as about 80°C

TABLE 1.8
Experimental Resins for High Solids Coatings
Recommended Formulations
and Their Properties

Experimental Resin	CLR-400	CLR-350	CLR-300
Resins Properties			
Solids	90.6% w	90.6% w	90.6% w
Viscosity, Gardner-Holdt	Z-6	Z-5	Z-4
Viscosity at 80% w solids, G-H	X+	W	U+
Solvent	MIBK	MIBK	MIBK
Color, Gardner	5	5	4
Coating Formulation			
Experimental resin (90.6% w solids)		74.6% w	
Cymel 303		16.9	
Phosphoric acid solution			
(10% in isopropyl alcohol)		8.5	
		100.0% w	
Epoxy resin/melamine resin		80/20	
Phosphoric acid		1 part/100 of resin solids	
Solids by weight, calculated		85% w	
Solids by volume, calculated		70% v	
Coating viscosity, poise			
at 25°C	31.9	23.4	20.1
at 66°C	2.2	2.0	1.9
Film Properties			
Film thickness	0.2 mils	0.2 mils	0.2 mils
Application		Drawdown via wire-wound bar at room temperature	
Bake schedule		10 min at 200°C	
MEK resistance[a] (double rubs)	90	60	40
Flexibility[b]	Pass 1T	Pass 1T	Pass 1T
Steam processing[c]	Pass	Pass	Pass

[a]Number of double rubs to failure.

[b]Number of thicknesses of substrate in radius of bend.

[c]90 min at 15 psig steam.

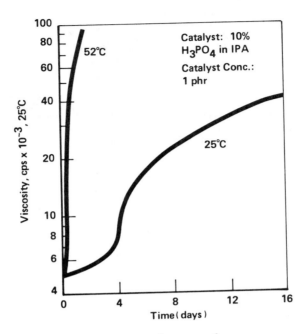

FIGURE 1.6   Effect of storage at 25 °C and 52 °C
             phosphoric acid catalyzed CLR-400/Cymel 303
             coating formulations in cellusolve acetate.

in 30 min. Typical formulations at 70% w solids using a melamine-
formaldehyde crosslinker are given in Table 1.9, and the excellent storage
stability at 55 °C is illustrated in Figure 1.8.

   Film performance characteristics including salt-spray data obtained
on resins of two different molecular weights cured with both types of aminoplast crosslinkers are compared in Table 1.10 with those obtained from a
commercial automotive primer. Salt-spray resistance is excellent, and
after 400 h was found to be superior to the commercial automotive primer
used as a control in this work. Film performance properties, such as
impact resistance and flexibility can be varied within limits by the choice
of epoxy resin molecular weight, and resin to crosslinker ratio. The
combination of a higher molecular weight epoxy resin (CLR 400) and/or a
higher ratio of epoxy resin to crosslinker (Table 1.6) appears to give cured
coatings with the optimum balance of properties.

   The resins described above as already indicated are based on conventional bisphenol-A epoxy resins, and were developed to supplement the
higher molecular weight solid epoxy resins widely used currently in baking
finishes. Now a new and unique class of experimental epoxy resins are

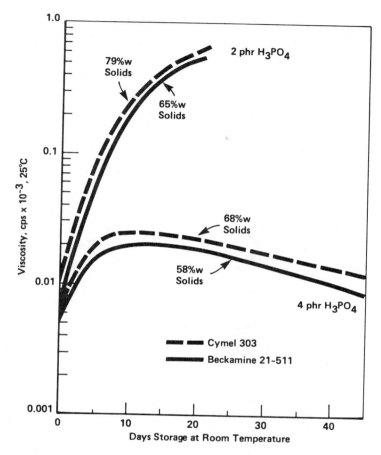

FIGURE 1.7   High solids coatings stability of unpigmented low bake systems. Coating formulation CLR-400/amino resin - 60/40 H_3PO_4 added as 10% IPA solution.

under development, which will add even greater dimensions to formulating epoxy coatings for a broad range of applications and that offer tremendous potential for high solids ambient cure epoxy coatings. The lower viscosity of these new saturated bisphenol-A epoxy resins compared with conventional bisphenol-A type epoxy resins not only permits coatings to be formulated at higher solids, but can also result in lower viscosity products when they are further reacted using typical epoxy chemistry. In addition, these products have improved exterior performance over unsaturated aromatic type epoxy resins while retaining

TABLE 1.9

High Solids Coatings Experimental Low Bake
Automotive Primer Formulation

Binder	
CLR-400/CYMEL 303	70/30
Phosphoric acid[a]	4 phr
Pigmentation/Pigments	
Barytes (No. 1)	60
Iron oxide (RO-5097)	20
Clay (ASP-400)	10
Talc (Asbestine 325)	10
	100
Nuosperse 657	0.5
Pigment/binder	2/1, 3/1, 4/1
Pigment volume concentration, %v	36, 46, 53
Weight solids, %w	70, 70, 70
Solvent	
MIBK/xylene/Oxitol® Glycol ether	1/1/1

[a]Added as a 10% w solution in butyl alcohol.

TABLE 1.10

Summary of Properties of Low Bake
Automotive-Type Primers
Crosslinking Resin: CYMEL 303

CRL Resin	CLR Resin Crosslinker	Flexibility Pass	Impact Resistance, in.-lb		Adhesion	
			Direct	Reverse	Primer	Topcoat
350	60/40	3/16	50	30	10	10
350	70/30	1/4	160	90	10	10
350	80/20	3/16	160	100	10	10
400	60/40	1/4	—	—	10	10
400	70/30	3/16	60	30	10	10
400	80/20	3/16	120	40	10	10
Control	—	3/16	160	140	10	10

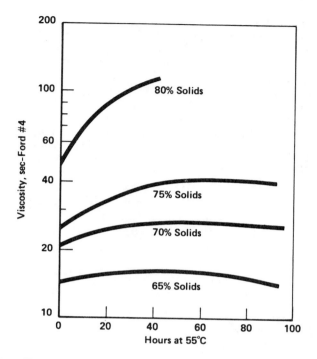

FIGURE 1.8   Experimental low-bake automotive primer viscosity
               stability at 55°C.

epoxy-like properties.[*]   Consequently, these new resins have great utility
in premium coating systems which demand the weatherability of acrylics or
urethanes, but require the performance of an epoxy. Typical properties of
these new low viscosity weatherable epoxy resins are given below:

Weatherable Epoxy Resins

Weight per epoxide (WPE)	232–238
Viscosity, poise, 25°C	20–25
Color, Gardner	1–2
Weight per U.S. gal. lb	9.08

---

[*]These new resins have already been discussed briefly in the water-borne
section above.

    These new low-viscosity, weatherable epoxy resins can be directly substituted for the bisphenol-A epoxy resin in the epoxy/dimer acid binders. Automotive primers based on the weatherable epoxy/dimer acid binder have shown no evidence of chalking or loss of intercoat adhesion under a clear acrylic topcoat after one-year outdoor exposure. The bisphenol-A epoxy control was chalking severely and had very poor intercoat adhesion after the same exposure.

    Also these new materials have been formulated directly into high solids baking finishes with outstanding gloss and color retention. A typical formulation is given in Table 1.11. Florida exposure results with this type of system are compared in Figure 1.9 along with the results obtained

TABLE 1.11

Aminoplast Cured Weatherable Epoxy Resin
White Baking Enamel

	Pounds	Gallons
Precatalyzed Epoxy Resin		
Weatherable epoxy resin	421	46.35
Phosphoric acid solution (10% w in IPA)	168	24.16
	589	70.51
Roller Mill Grind		
Precatalyzed epoxy binder[a]	589	70.51
TiO$_2$ (Ti-Pure R-960)[b]	982	29.49
	1571	100.00
Enamel Letdown		
Roller mill grind	694	44.18
Cymel 325 (80% N.V.)[c]	200	21.47
FC-430[d]	2	0.22
Cellosolve acetate/xylene (1/1)	261	34.17
	1157	100.00
Enamel Constants		
Spray viscosity, #4 Ford cup-sec.	21	
Spray solids, % v	49.6	
Bake Schedule: 150°C for 30 min		

[a]Allow acid to sweat in for 3 days prior to pigmenting.

[b]E. I. DuPont de Nemours Co. Inc.

[c]American Cyanamid Co.

[d]The 3M Company.

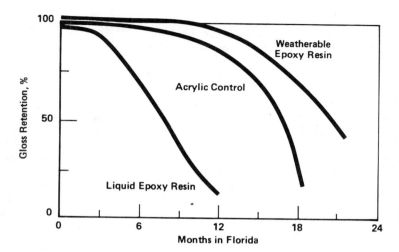

FIGURE 1.9   Florida exposure data on the weatherable epoxy resin
baking enamel and selected controls.

from a corresponding coating prepared with a conventional bisphenol-A
liquid epoxy resin and a commercially-available, baked-acrylic system.
Also these new weatherable epoxy resins showed nonchalking and nonyellowing characteristics superior to both the acrylic control and to the
corresponding system based on the conventional liquid epoxy resin (Table
1.12).

Higher Solids Ambient-Cure Finishes

Ambient-cure coal tar epoxy coatings and ketimine-cured coatings
based on conventional liquid bisphenol-A epoxides have for many years
offered high solids coatings with greater than 70% volume solids. These
systems, however, have a number of shortcomings and have achieved only
limited application. New epoxy resin and epoxy resin curing agent technology is now emerging that will add greater dimensions to formulating
epoxy coatings for a broad range of applications and that will offer the
potential for higher solids coatings.
The low viscosity of the new experimental weatherable resin already
mentioned offers the potential for higher solids coatings than now possible
from conventional bisphenol-A epoxides. Viscosity data for pigmented
coatings formulated from this resin and two new polyamide curing agents
designed specifically for use with the weatherable resin are given in
Figure 1.10. Substantially higher solids ambient-cure coatings can be

TABLE 1.12
Nonchalking and Nonyellowing Characteristics of
the Weatherable Epoxy Resin Baking Enamel
and Selected Controls

Months in Florida	Yellowing (Δ YI)			Chalking		
	Weatherable Epoxy Resin	Acrylic	Liquid Epoxy Resin	Weatherable Epoxy Resin	Acrylic	Liquid Epoxy Resin
6	0.2	-1.1	4.2	10	10	10
12	0.5	0.9	5.7	10	10	10
18	1.7	1.8	9.6	10	10	-
21	-	-	-	10	10	6

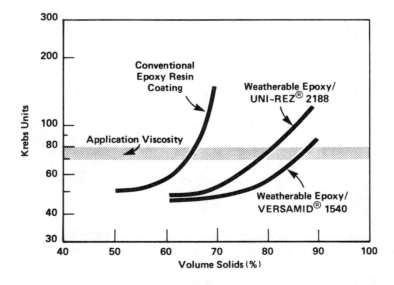

FIGURE 1.10   Krebs units vs. percent volume solids selected two-
package weatherable coatings.

obtained from these systems than from the more conventional epoxy coatings with volume solids of 70% or greater possible at the application range of 70% or greater possible at the application range of 70 to 80 Krebs units.

Additionally, encouraging accelerated tests (both Weather-ometer and cyclic ultraviolet weathering tests), as well as preliminary outdoor weathering results, have been obtained with this weatherable epoxy resin in ambient-cure coatings using either of two polyamide curing agents, Versamid 1540 (Henkel Corporation) or Uni-Rez 2188 (Union-Camp Corporation). The weathering characteristics and film performance of the two solvent-borne weatherable polyamide cured systems have been examined extensively. The formulations for these experimental coatings are given in Table 1.13 and the outstanding gloss retention obtained with these ambient-cure weaterable epoxy resin coatings is clearly evident from the results shown in Figure 1.11 after 600 hours exposure in an ultraviolet cyclic tester (QUV). This improved weatherability over conventional bisphenol-A epoxy resin based coatings, which accelerated tests appear to indicate approaches that of polyester urethanes, is available without significant sacrifice of traditional epoxy coating film performance. Enamel characteristics and film properties obtained with these new resins suggest they show outstanding potential as a replacement for urethane coatings in industrial maintenance, marine, aircraft, and automotive refinish.

This resin also is a versatile building block for higher solids coating resins, since it can be reacted using typical epoxy chemistry to give products of lower viscosity than obtained from conventional liquid epoxy resins. Lower viscosity solid type resins, for example, are available by reaction of the weatherable epoxy resin with bisphenol-A. These solid resins can be further esterified to low viscosity air dry epoxy esters with unsaturated fatty acids. The weatherability of these solid resins will, however, be reduced depending on the amount of bisphenol-A used in the up-staging reaction. All of these products, as well as several other low viscosity resins for high solids coatings based on these new resins, are discussed in detail below.

A conventional solid epoxy resin with a weight per epoxide in the range of 900 will typically have solution viscosities (40% w in Butyl DIOXITOL® flycol ether at 25°C) of between 400 and 660 centipoises. A corresponding solid prepared from the weatherable resin and bisphenol-A will have a solution viscosity of about 140 centipoises. These resins will then permit formulation of higher solid coatings of the type based on solid epoxy resins which will cure with all the conventional crosslinkers normally used in these systems. Again, the gloss retention of coatings prepared with bisphenol-A modified weatherable epoxy resin will be decreased according to the amount of bisphenol-A used.

Using resorcinol in place of bisphenol-A to advance the weatherable resin, a higher molecular weight epoxy resin is obtained having improved gloss retention over a conventional solid epoxy resin or the corresponding

TABLE 1.13
Weatherable Epoxy Resin White Enamel Coatings
Solvent-Borne, Two Package, Ambient Cure

	A		B	
	lb	gal	lb	gal
Pigmented Base				
TiO$_2$ (Ti-Pure R-960)[a]	333.27	9.53	288.75	8.25
Bentone 27[b]	5.14	0.36	4.00	0.25
Methanol	3.08	0.47	3.14	0.48
Nuosperse 657[c]	1.03	0.13	1.19	0.15
Beetle 216-8[d]	5.14	0.60	4.03	0.47
Weatherable epoxy resin	236.58	26.06	204.94	22.57
	37.15	506.05	32.30	32.17
Curing Agent Component				
Versamid 1540	129.61	16.22	–	–
Uni-Rez 2188/75% w solids				
in n-Butanol	–	–	201.40	25.82
n-Butanol	314.61	46.63	284.62	41.98
	1028.61	100.00	992.07	100.00
Enamel Constants				
Pigment/binder	0.9/1		0.8/1	
Solids, % w	68.9		65.4	
Solids, % v	52.3		49.9	
Curing Agent, phr	55		74	
Viscosity, No. 4 Ford cup-sec	24		22	

[a]E. I. DuPont de Nemours Co., Inc.

[b]Bentone and methanol mixed prior to addition.

[c]Tenneco Corp.

[d]American Cyanamid Co.

weatherable resin up-staged with bisphenol-A. The gloss retention characteristics of the resorcinol advanced system cured with a melamine-formaldehyde crosslinking resin is shown in Figure 1.12 which also contrasts the performance of corresponding coatings based on a conventional solid epoxy resin and a bisphenol-A advanced weatherable epoxy resin.

In turn, the above solid resins are esterifiable with fatty acid using conventional esterification technology. The resulting esters have

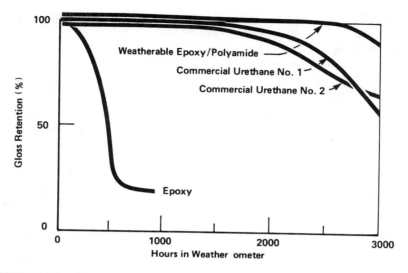

FIGURE 1.11   Gloss retention characteristics two-package formulation.

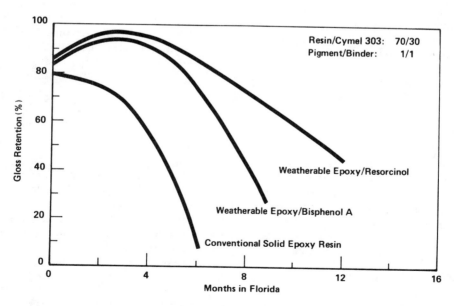

FIGURE 1.12   Florida exposure data solid weatherable epoxy resin.

viscosities significantly lower than those obtained from bisphenol-A solid epoxy resins, which permits formulation of epoxy ester type coatings with higher solids content. A classic D-4 type epoxy ester prepared from a conventional solid epoxy resin with a weight per epoxide of 841 and de- hydrated castor fatty acid had a solution viscosity of about 1500 cps (60% w in xylene). The viscosity of the ester prepared from a bisphenol-A mofified weatherable epoxy resin having a weight per epoxide of 864 was 160 cps (60% w in xylene). This ester was formulated into a white enamel topcoat at 58% v solids which had a paint viscosity of 68 Krebs units, whereas the corresponding conventional ester at 45.0% v solids had a paint viscosity of 73 Krebs units.

An ultraviolet (UV) curable diacrylate ester has also been prepared, and again a product of much lower viscosity is obtained than from a conven- tional liquid epoxy resin. Viscosities in the range of about 1000 poises are typical for the diacrylate of this resin, which is significantly lower than the 9000 poises for the BPA epoxide diacrylate.

CONCLUSION

Pressures for environmentally-acceptable and energy-efficient coat- ings as well as demands for even higher performance systems continue to challenge the resin chemist. This paper specifically addressed itself to recent advances being made in the area of new epoxy resin coatings, and covered some of the new resins, curing agents, and technology being developed to meet these challenges presented to us. Although compre- hensive coverage of all the recent developments in this area was not pos- sible in this single presentation, it is hoped that the new developments discussed here will demonstrate the versatility of epoxy resins as building blocks for the premium coatings of the future.

REFERENCES

1. Modern Plastics, 58, No. 1 (January 1980).
2. R. D. Jerabek, U. S. Patent 3,984,299 (PPG Industries).
3. J. F. Bosso, and M. Wismer, U. S. Patent 3,962,165 (PPG Industries).
4. J. F. Bosso, and M. Wismer, U. S. Patent 3,959,106 (PPG Industries).
5. J. F. Bosso, and M. Wismer, U. S. Patent 3,894,922 (PPG Industries).
6. a) W. Raudenbusch in R. S. Bauer (Ed.), Epoxy Resin Chemistry,
        American Chemical Society, Washington, D.C., 1979, p. 57;
    b) W. J. Van Westrenen, J. Oil Colour Chem. Assoc. 62, 246 (1979).
7. R. A. Allen, and L. W. Scott, U. S. Patent 4,098,744 (Shell Oil).
8. E. M. A. A. J. Van Acker, "Modern Developments in Formulating
    and Testing of Can Coatings," XVth Fatipec Congress, II, 45 (1980).

9. J. M. Evans and V. W. Ting, U. S. Patent 4,212,781 (SCM Corp.).
10. P. H. Martin and R. T. McFadden, U. S. Patent 3,719,629 (Dow Chemical).
11. R. S. Bauer, "Weatherable Epoxy Resins," 14th Western Coatings Societies Symposium, San Francisco, 1979.
12. K. F. Schimmel et al., U. S. Patent 4,126,596 (PPG Industries).
13. K. F. Schimmel et al., U. S. Patent 4,221,885 (PPG Industries).
14. J. D. Elmore and H. T. Dickma, "Design of Resins for Ambient Cure Water-Borne Epoxy Coatings," Water-Borne and Higher Solids Coatings Symposium, New Orleans, 1980.
15. R. S. Bauer and J. A. Lopez, U. S. Patent 4,119,595 (Shell Oil).

PHOTOINITIATED CATIONIC POLYMERIZATION
OF COATINGS

James V. Crivello

General Electric Corporate Research and Development Center
Schenectady, New York

ABSTRACT

The synthesis, photosensitivity, and mechanism of photolysis of triarylsulfonium salt photoinitiators is discussed. Data are presented describing the effects of structure, light intensity, presence of oxygen, and moisture on the use of these compounds in the photoinitiated polymerization of epoxides. The use of photosensitizers to increase the overall efficiency of triarylsulfonium salt-photoinitiated cationic polymerization is described.

INTRODUCTION

Background

In recent years, there has been an increasing interest in the use of photopolymerizations for many industrial coating and printing processes. Because of their high cure and application speeds, essentially pollution-free operation, very low energy requirements and generally excellent properties, coatings prepared by photopolymerizations have made a substantial impact on the wood coating, metal decorating, and printing industries. While the bulk of the current research effort continues to be directed toward photoinduced radical polymerizations, it is well-recognized that ionic photopolymerizations also hold considerable promise in many

application areas. Photoindiced cationic polymerizations are particularly
attractive because of the wide breadth of chemical and physical properties
which can potentially be realized through the polymerization of such mono-
meric substrates as electron rich olefins, lactones, epoxides, cyclic
ethers, sulfides, silicones, and acetals as well as numerous other hetero-
cyclic compounds. Further, photoinitiated cationic polymerizations have
the important commercial advantage that they are not inhibited by oxygen
and thus may be carried out in air without the need for blanketing with an
inert atmosphere to achieve rapid and complete polymerization [1].

In addition to the above enumerated practical considerations, the
design of photoinitiator molecules capable of initiating cationic polymerization represents a significant scientific challenge in the area of polymer
chemistry. It is symptomatic of the complexity of this problem that although photoinitiators for radical polymerication have been known and
employed for over 40 years, it has only been within the last 10 years that
significant progress has been made toward the development of analogous
photoinitiators for cationic polymerization.

Criteria

Before going into specific details of these investigations, it is important to consider the necessary and practical criteria which must be taken
into account in the design, synthesis, and evaluation of photoinitiators for
cationic polymerization. Most importantly, the compound must possess
chromophors which permit the absorption of UV or visible light and as a
result, undergo some photochemical transformation which generates a
species capable of initiating cationic polymerization. Before photolysis,
the ideal photoinitiator should be indefinitely stable by itself and when dissolved in highly reactive monomers. When irradiated, it should generate
the initiating species with a high-quantum efficiency and without the simul-
taneous liberation of by-products which inhibit or retard polymerization.

The first breakthrough in the development of cationic photoinitiators
followed the discovery by two groups [2-7] that aryldiazonium salts could
be employed as photoinitiators for the polymerization of epoxides (equations (1)-(3)).

$$\overset{+}{Ar-N}\equiv N\ \overset{-}{BF}_4 \xrightarrow{\ h\nu\ } Ar-F + N_2 + BF_3 \tag{1}$$

$$BF_3 + CH_2 - \underset{\underset{R}{|}}{CH} \longrightarrow \overset{-}{BF_3}O-CH_2-\overset{+}{\underset{\underset{R}{|}}{CH}} \tag{2}$$

$$\overset{-}{BF_3})-CH_2-\overset{+}{\underset{\underset{R}{|}}{CH}} + n\ CH_2-\underset{\underset{R}{|}}{CH} \longrightarrow \overset{-}{BF_3})\!\!+\!\!CH_2-\underset{\underset{R}{|}}{CH}-O\!\!\rightarrow\!\!_n CH_2-\overset{+}{\underset{\underset{R}{|}}{CH}} \tag{3}$$

Despite the apparent wide applicability of aryldiazonium salts, particularly in epoxy polymerizations, there are several inherent drawbacks of these salts which limit their utility in a number of practical applications. One of the by-products produced in the photolysis of diazonium salts is nitrogen gas. The presence of nitrogen causes bubbles and pinholes in cured films thicker than a few tenths of a mil. Thus, potential uses are restricted to very thin film applications such as container coatings and photoresists. Another problem arises from the poor thermal stability of aryldiazonium salt photoinitiators. Currently, even with stabilizers, aryldiazonium salt-epoxy systems have a maximum pot life of about 2 wks at room temperature and for this reason, are generally regarded as two-part systems [8].

Our research group at General Electric, along with similar groups at the 3M Company and ICI have been successful in their attempts to develop photoinitiators which overcome the difficiencies of the aryldiazonium salts, namely, diaryliodonium [9-13], triarylsulfonium [14-17], triarylselenonium [18], dialkylphenacylsulfonium [19] and dialkylhydroxyphenylsulfonium salts [20], which are highly efficient photoinitiators for the polymerization of epoxides, lactones, cyclic ethers, cyclic sulfides, cyclic acetals, and vinyl compounds.

The scope of this present paper will be confined to an examination of only one of these new classes of photoinitiators, triarylsulfonium salts. These photoinitiators are among the most photoactive, are easily prepared and since they are commercially available, considerable data exists describing their use in a variety of practical coating applications.

TRIARYLSULFONIUM SALTS

Triarylsulfonium salts having the general structure

$$Ar'-\overset{\overset{\displaystyle Ar}{|}}{\underset{\underset{\displaystyle Ar''}{|}}{S^+}} X^-$$

are a class of very stable crystalline photosensitive compounds whose preparation and characterization have been reported in the literature as early as 1891 [21]. Only those triarylsulfonium salts bearing nonnucleophilic anions of the type $X^- = ClO_4^-$, BF_4^-, PF_6^-, SbF_6^- are useful as photoinitiators of cationic polymerization.

Synthesis

A considerable number of synthetic routes have been developed for the preparation of symmetrical and unsymmetrical triarylsulfonium salts.

The most important of these methods will be briefly summarized here.

Complex mixtures of triarylsulfonium salts can be prepared simply by the direct condensation of aromatic hydrocarbons with sulfur dichloride or sulfur monochloride in the presence of a Lewis acid followed by chlorination and further condensation (equation (4)) [22].

$$6ArH + S_2Cl_2 \xrightarrow[\text{or } SnCl_4]{AlCl_3} \xrightarrow{3Cl_2} 2Ar_3S^+Cl^- + 6HCl \qquad (4)$$

The process shown in equation (4) is practiced on a commercial scale for $Ar = C_6H_5$ and provides a readily available source of these triarylsulfonium salts.

The Friedel-Crafts reaction of 2, 6-disubstituted phenols with thionyl chloride in the presence of aluminum chloride produces good yields of well-characterized crystalline salts [23]. However, it is limited to phenols as substrates and cannot be applied to other more deactivated compounds.

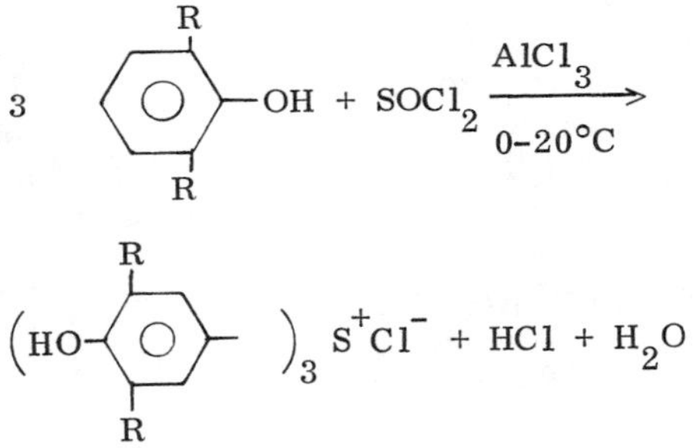

(5)

Another useful synthetic method is based on sulfilimine compounds. These intermediates can be prepared by various routes, however, the method described by Oae et al [24] and shown below is the best in terms of yield.

$$Ar_2S + CH_3\text{-}C_6H_4\text{-}SO_2\text{-}N\begin{matrix}Na\\Cl\end{matrix} \xrightarrow[H^+]{CH_3OH}$$

$$NaCl + Ar_2S{=}N\text{-}SO_2\text{-}C_6H_4\text{-}CH_3 \qquad (6)$$

Sulfilimines can be converted to sulfonium salts by condensing them with aryl hydrocarbons in the presence of aluminum chloride [25] (equation (7)).

$$\begin{array}{c} Ar \\ \diagdown \\ \diagup \\ Ar \end{array} S{=}N{-}Ts \;+\; Ar'H \xrightarrow[\substack{\Delta \\ 2\,h}]{AlCl_3} \xrightarrow[H_2O]{HX}$$

$$\begin{array}{c} Ar \\ \diagdown \\ \diagup \\ Ar \end{array} \overset{+}{S}{-}Ar'\;X^- \;+\; Ts{-}NH_2 \qquad\qquad (7)$$

This latter method seems to be applicable only to hydrocarbons that undergo facile electrophilic substitution; i.e., toluene, xylene, and mesitylene.

A general route to the synthesis of triarylsulfonium salts involves the condensation of diarylsulfoxides with Grignard reagents [26] [equation (8)].

$$\begin{array}{c} Ar \\ \diagdown \\ \diagup \\ Ar \end{array} S \longrightarrow O \;+\; Ar'{-}MgX \xrightarrow[\Delta]{C_6H_6} \xrightarrow{HX}$$

$$\begin{array}{c} Ar \\ \diagdown \\ \diagup \\ Ar \end{array} \overset{+}{S}{-}Ar'\;X^- \;+\; MgXOH \qquad\qquad (8)$$

All of the above methods of preparation give rise to triarylsulfonium salts with halide counterions which are inactive as photoinitiators for cationic polymerization. Therefore, these salts must be converted to the corresponding salts in which the anion $\underline{X}^-$ is nonucleophilic. This conversion may be accomplished using either of the two methods shown in equation (9).

$$Ar_3\overset{+}{S}Cl^- \xrightarrow{AgOH} Ar_3\overset{+}{S}OH^- \;+\; AgCl$$

$$NaX \searrow \qquad\qquad \downarrow HX$$

$$Ar_3\overset{+}{S}X^- \qquad\qquad (9)$$

In Table 2.1 are the structures, melting points, and ultraviolet spectral characteristics of some typical triarylsulfonium salt photoinitiators. These compounds are generally white crystalline salts indefinitely stable in the absence of light. In fact, although these salts are highly photosensitive, they display a surprising degree of thermal stability. Thermogravimetric analysis shows that thermal decomposition does not proceed until nearly $350°C$ in air. Triarylsulfonium salts are readily soluble in a wide variety of nitriles, chlorinated hydrocarbons, ketones, alcohols, and particularly important, in most cationically polymerizable monomers.

TABLE 2.1
Triarylsulfonium Salts

Cation	Anion	Mp (°C)	λ_{max} (εmax)(MeOH)	Elemental Analysis		
				C	**H**	**S**
1. $[C_6H_5]_3 S^+$	BF_4^-	191-193	230 nm (17,500)	calc. 61.71 / fnd. 62.00	4.28 / 4.31	9.14 / 9.33
2. $[C_6H_5]_3 S^+$	AsF_6^-	195-197	230 nm (17,500)	calc. 47.68 / fnd. 47.78	3.31 / 3.41	7.06 / 7.06
3. CH_3–C_6H_4–S^+–$[C_6H_5]_2$	PF_6^-	133-136	237 nm (20,400) / 249 nm (19,700)	calc. 56.90 / fnd. 57.05	4.96 / 5.03	6.90 / 7.09
4. $[CH_3O$–$C_6H_4]_3 S^+$	AsF_6^-	—	225 nm (21,740) / 280 nm (10,100)	calc. 46.49 / fnd. 46.70	3.88 / 3.98	5.90 / 6.00
5. $[CH_3$–$C_6H_4]_3 S^+$	BF_4^-	162-165	243 nm (24,700) / 278 nm (4,900)	calc. 64.28 / fnd. 64.32	5.35 / 5.41	8.16 / 7.91
6. $(CH_3)_2C_6H_3$–S^+–$[C_6H_5]_2$	AsF_6^-	111-112	275 nm (42,100) / 287 nm (36,800) / 307 nm (24,000)	calc. 50.00 / fnd. 50.10	3.95 / 4.01	6.67 / 7.23
7. $[HO$–$(CH_3)_2C_6H_2]_3 S^+$	AsF_6^-	245-251	263 nm (25,200) / 280 nm (22,400) / 316 nm (7,700)	calc. 49.39 / fnd. 49.39	4.62 / 4.59	5.48 / 5.55
8. (thianthrenium/phenyl dibenzothiophenium structure)	BF_4^-	168-169	277 nm (25,000) / 232 nm (3,100)	calc. 62.99 / fnd. 62.81	4.14 / 4.10	8.84 / 8.96
9. (phenoxathiin sulfonium with phenyl)	AsF_6^-	165-168	238 nm (20,600) / 292 nm (5,200)	calc. 46.35 / fnd. 46.15	2.78 / 2.76	6.86 / 6.95
10. (phenoxathiin sulfonium with 4-Cl-phenyl)	AsF_6^-	183-187	238 nm (22,600)	calc. 43.20 / fnd. 43.11	2.40 / 2.45	6.40 / 6.31

Photolysis and Mechanistic Studies

Although triarylsulfonium salts are highly thermally stable, they undergo rapid photolysis when irradiated at wavelengths from 200-300 nm. Under these conditions, efficient (Φ_{313} = 0.17-0.19) homolytic rupture of one of the carbon-sulfur bonds results. The following mechanism has been proposed for this photolysis [14-27]:

$$Ar_3S^+X^- \longrightarrow [Ar_3S^+X^-]^* \qquad [a]$$

$$[Ar_3S^+X^-]^* \longrightarrow Ar_2S^{+\cdot} + Ar\cdot + X^- \qquad [b]$$

$$Ar_2S^{+\cdot} + YH \longrightarrow Ar_2S^+-H + Y\cdot \qquad [c]$$

$$Ar_2S^+-H \longrightarrow Ar_2S + H^+ \qquad [d]$$

$$2Ar\cdot \longrightarrow Ar-Ar \qquad [e]$$

$$Ar\cdot + YH \longrightarrow ArH + Y\cdot \qquad [f] \qquad (10)$$

Interaction of the radical-cation $Ar_2S^{+\cdot}$ with the solvent or monomer, YH, in step [c] results in the release of a proton in step [d] and the resultant formation of the strong acid HX. In addition to hydrogen abstraction, $Ar_2S^{+\cdot}$ may also undergo electrophilic attack on the monomer or other impurities in the system to generate acids. Brønsted acids such as HBF_4, $HAsF_6$, and $HSbF_6$ are well-known very reactive initiators of cationic polymerization [28].

Although the process shown in equation (10) is radical in nature, the failure of radical inhibitors to affect the rate suggests that a radical induced chain process is either not involved or that the kinetic chain is very short. In addition, the photolysis of triarylsulfonium salts is not quenched by molecular oxygen nor has it been found possible to quench their photolysis using other common triplet quenchers.

Other evidence in support of the mechanism shown in equation (10) includes the following: [a] the proposed primary photolysis step is consistent with the results of earlier studies on the photochemistry [29] and electrochemistry [30] of triarylsulfonium salts as well as the analogous diaryliodonium salts [31], [b] the quantum yield of diarylsulfide formation was found to be independent of the counterion (X⁻) [14], and [c] initiation of cationic polymerization is consistent with the formation of a long-lived

species such as the protonic acid, HX, since preirradiation of a triarylsulfonium salt in the absence of a monomer followed by introduction of the
monomer results in rapid polymerization [15].

The rate at which various triarylsulfonium salts undergo photolysis
is a function of the structure of the cationic portion of the molecule as
shown in Figure 2.1. Manipulation of the absorption characteristics of
these compounds through the introduction of various substituents can exert
a profound effect on their apparent photosensitivity. At the same time,
Figure 2.2 shows that the structure of the anion has no effect on the rate
of photolysis of tris-(4-methylphenyl)sulfonium salts subjected to the same
conditions of irradiation.

While the photolysis of triarylsulfonium salts has been observed to
be independent of temperature, there is a marked dependence of the
photolysis rates on both the intensity and wavelength of irradiation. Figure
2.3 shows the relationships between light intensity and photolysis rate for
a typical triarylsulfonium salt. As the curves indicate, a direct proportional relationship exists between the intensity and the rate of photolysis.
From a practical standpoint, these results suggest that the speed of a
cationic photopolymerization may be increased twofold simply by doubling
the number of lamps. In practice this has been observed to be the case.

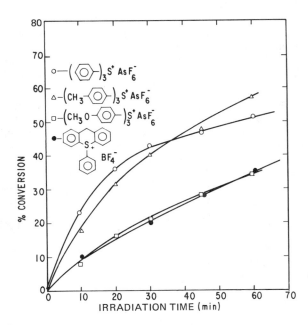

FIGURE 2.1   Photolysis of various triarylsulfonium salts (0.07M in
acetone).

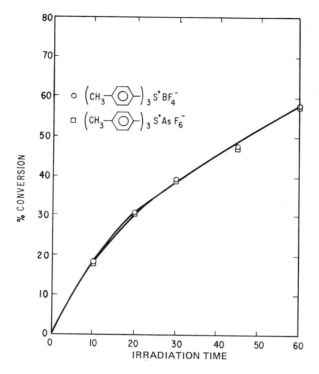

FIGURE 2.2   NMR study of the photolysis of tris-(4-methylphenyl)-
sulfonium salts (0.35M) in acetone-d_6.

Photoinitiated Cationic Polymerization Using Triarylsulfonium Salts

The chief mechanism by which triarylsulfonium salts photoinitiate the polymerization of cyclic ether and olefinic monomers is given in equations (11)-(13).

Photolysis

$$Ar_3S^+X^- \xrightarrow[\text{solvent-H}]{h\nu} Ar_2S + Ar\text{-}H + HX_{(s)} \tag{11}$$

Initiation

$$M + HX_{(s)} \rightleftharpoons HM^+X^- \tag{12}$$

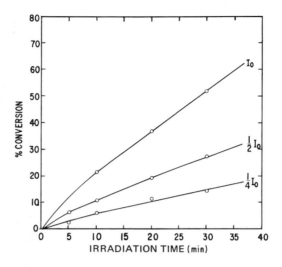

FIGURE 2.3   Photolysis of triphenylsulfonium hexafluoroarsenate in
acetone using 350 nm light at various intensities.

Propagation

$$\text{HM}^+ \text{X}^- + n\text{M} \longrightarrow \text{H(M)}_n \text{M}^+ \text{X}^- \qquad (13)$$

The first step involves the photogeneration of the strong protonic
acid HX. Cationic polymerization occurs in subsequent dark (nonphoto-
chemical) steps. Initiation results from the direct protonation of the
monomer by HX to form a carbenium or onium species. Spectroscopic
analyses on polymers generated from triarylsulfonium salt photoinitiators
confirm the absence of aromatic end groups originating from fragments oı
the photoinitiators. This is additional evidence in support of a protonic
acid as the major initiator. In the chain growth step [equation (13)] the
active center propagates by the stepwise addition of monomer molecules.
    Although the character of the anion present in a triarylsulfonium salt
plays no role in determining its rate of photolysis, it does exert a considerable influence on the rate and extent of a cationic polymerization. At
one extreme, triarylsulfonium salts bearing nucleophilic anions such as
I^-, Cl^-, Br^- and CH_3CO_2^- compete so successfully with all cationically
polymerizable monomers for the active cationic species that polymeriza-
tion invariably fails. When anions of intermediate nucleophilicity such as
HSO_4^-, FSO_3^-, NO_3^- and CF_3CO_2^- are employed, successful polymerizations of certain monomers, e.g., vinyl ethers, are observed while still
others fail to polymerize at measurable rates. On the other hand,

triarylsulfonium salts containing such nonnucleophilic anions as ClO_4^-, $CF_3SO_3^-$, BF_4^-, AsF_6^-, PF_6^- and SbF_6^- are capable of polymerizing virtually all known types of cationically polymerizable monomers. Included among these monomers are epoxides, cyclic ethers, mono- and polyfunctional vinyl compounds, spiroesters, spirocarbonates, and cyclic siloxanes. Because of the importance of epoxy resins in the coatings industry—last year over 100 million lbs. were used for this purpose—this portion of the paper will deal exclusively with a characterization of the triarylsulfonium salt photoinitiated cationic polymerization of these resins.

Figure 2.4 shows a study of the photoinitiated bulk polymerization of styrene oxide using equimolar amounts of triphenylsulfonium salts bearing the BF_4^-, PF_6^- and AsF_6^- anions. Rapid, nearly explosive polymerization was noted in the case of the triphenylsulfonium SbF_6^- salt. Since the photolysis rates for the four salts are the same in all cases, identical amounts of the corresponding Brønsted acids are produced per unit irradiation time. The order of reactivity of triphenylsulfonium salts in the polymerization of styrene oxide is therefore: $SbF_6^- > AsF_6^- > PF_6^- > BF_4^-$.

The above model studies with monofunctional epoxy compounds give considerable insight into the reactivity of various triarylsulfonium salt photoinitiators. However, in terms of coatings, one is primarily interested in the reactivity of these photoinitiators in di- and multifunctional

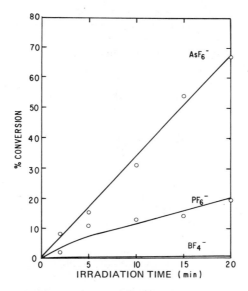

FIGURE 2.4   Photoinitiated polymerization of styrene oxide 0.02 M $(C_6H_5)_3S^+X^-$ as photoinitiators.

TABLE 2.2
UV Cure of ERL 4221 Using Various
Triphenylsulfonium Salts

Photo Photoinitiator	Cure Rate (ft/min)[a]
$(C_6H_5)_3S^+ BF_4^-$	195
$(C_6H_5)_3S^+ PF_6^-$	228
$(C_6H_5)_3S^+ AsF_6^-$	248
$(C_6H_5)_3S^+ SbF_6^-$	280

[a]1.0 mil films cured using three 200 W/in.
Hanovia Hg arc lamps aligned parallel to
direction of travel of the conveyor.

epoxy systems. Determination of the cure rates required to produce tack-free coatings is a generally accepted industry standard for the reactivity of UV cured coatings. In Table 2.2 are given the cure rates in ft/min which were determined for the cycloaliphatic bisepoxide, 3,4-epoxycyclohexyl-methyl-3',4'-epoxycyclohexane carboxylate (ERL 4221) using 1 mol % of triphenylsulfonium BF_4^-, PF_6^-, AsF_6^- and SbF_6^- salts.

A more quantitative measure of the degree of cure as well as the cure rate in UV cure systems can be obtained with the aid of a specially designed differential scanning calorimeter. This device was developed at General Electric [32] and described by the present author in previous publications [33-35]. Figure 2.5 shows the type of data which can be obtained using this device. When the shutter is opened allowing UV light to strike the sample, immediate exothermic reaction ensues which is recorded as a peak by the strip chart recorder. The time interval from when the shutter is opened to the peak of the exothermic curve is one measure of the reactivity of the polymerization system under study: the shorter the time interval between these two points, the more reactive the system. In addition, the heats of reaction, which correspond to the areas under the curves are a quantitative measure of the extent of polymerization. Comparing the various curves in Figure 2.5 leads again to the conclusion that the character of the anion present in the photoinitiator has a marked effect on both the rate and extent of polymerication.

Other important rate determining factors in the photoinitiated polymerization of epoxies are the concentration of the photoinitiator, the

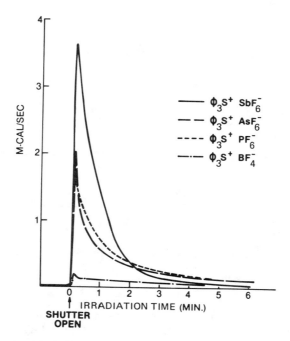

FIGURE 2.5   Photopolymerization of ERL 4221 using 1.5 M % $(C_6H_5)_3S^+X^-$ salts.

wavelength of the irradiating light and its intensity and the temperature. The influence of these factors can be conveniently determined using the modified differential scanning calorimeter (DSC).   Figure 2.6 shows a plot of the effect of concentration of $(C_6H_5)_3S^+ AsF_6^-$ on the rate of cure of of ERL 4221.   The optimum cure rates are obtained at concentrations of 2%-3% of this photoinitiator.   Further increase in the photoinitiator level does not produce a corresponding increase in the cure rate, possibly due to light screening effects by the triarylsulfonium salt itself or its photolysis products.

   A DSC study of the effect of light intensity on the cure rate of ERL 4221 containing 2% by weight $(C_6H_5)_3S^+ AsF_6^-$ is displayed in Figure 2.7. In this figure, I_0 is the incident light intensity delivered by a GE H3T7 medium pressure mercury arc lamp ballasted at 200 Watts/inch of arc length and positioned at 20 cm from the samples.   As the curve shows, the cure times approach limiting values at both high and low light levels.   At high light intensities, the system is limited by the absorption characteristics of photoinitiator, while at very low intensities, there appears to be some type of threshold or inhibition effect.

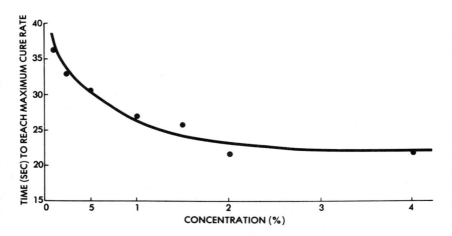

FIGURE 2.6   Concentration dependency of the cure rate of ERL 4221 on %
$(C_6H_5)_3S^+ AsF_6^-$.

        Figure 2.8 shows the effect of temperature on the polymerization of
ERL 4221 using 2% $(C_6H_5)_3S^+ PF_6^-$ as the photoinitiator. By increasing
the temperature from 25°C to 85°C, the time required to reach the maximum cure rate has been decreased by one-half. In all photoinitiated
cationic polymerizations employing triarylsulfonium salt photoinitiators,
cure at the highest temperature the substrate will allow, will give the
highest cure rates. Obviously, certain trade-offs must be made between
the cure temperature and loss of the monomer through volatilization.
        Photoinitiated cationic polymerizations of epoxides are unaffected by
the presence of atmospheric oxygen. There is, therefore, no need to
blanket them with nitrogen. Water and other hydroxyl containing impurities
can be tolerated in amounts of 1%-2% in most epoxy monomers without
seriously affecting their polymerizability. However, the presence of water
does change considerably both the rate and extent of polymerization of
epoxy monomers. Figure 2.9 shows the DSC curves for the photopolymerization of both dry ERL 4221 and the same epoxy resin containing 1%-2% by
weight water, using $(C_6H_5)_3S^+ SbF_6^-$ as the photoinitiator. The presence
of a second peak in the DSC curve of the wet ERL 4221 indicates the presence of another competing chemical process taking place. Basic materials
present in UV curable formulations as either impurities, additives, or
fillers can inhibit the polymerization and should be avoided. UV-opaque
fillers can be tolerated to surprising levels. In Figure 2.10 is shown a
plot of the heat of reaction, ΔH, against the level of TiO_2 (anatase) in a
formulation containing ERL 4221. The higher the TiO_2 concentration
present, the lower the extent of polymerization which takes place due to the

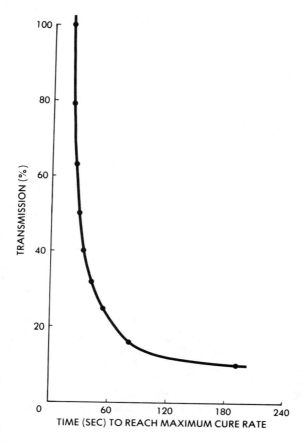

FIGURE 2.7   Effect of light intensity on the cure rate of ERL 4221 with
2% $(C_6H_5)_3S^+ AsF_6^-$.

screening effect of the pigment. Nevertheless, even at TiO_2 levels of 12%,
hard, tack-free coatings can be obtained provided that the coatings are
reasonably thin (< 1 mil).

Photosensitization

Although triarylsulfonium salts absorb strongly at wavelengths near
250 nm, their absorption at longer wavelengths is comparatively low.
Introduction of simple substituents on the aryl ring does not markedly alter
their spectral characteristics. The poor absorbtivity of triarylsulfonium

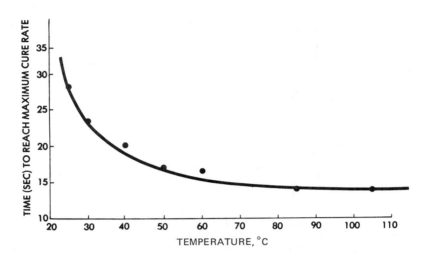

FIGURE 2.8   Effect of temperature on the cure rate of ERL 4221 with 2%
$(C_6H_5)_3S^+ PF_6^-$.

salts in the 300 to 450 nm region severely limits their efficiency of light
utilization at 313, 366, 405, and 436 nm where medium and high-pressure
mercury arc lamps provide a substantial portion of their emission.
Recently, the photosensitization of triarylsulfonium salt photolysis by
dyes has been observed [34-36] making it possible to photoinitiate cationic
polymerization with both long wavelength UV and visible light. Inclusion
of a photosensitizer into a coating formulation can dramatically reduce the
cure time as shown in Table 2.3 by greatly increasing the efficiency of
light utilization.

CONCLUSION

Over the past ten years, the development of triarylsulfonium salt
and other cationic photoinitiators has moved from the realm of speculative
investigation to the point today at which they are being employed in numerous commercial applications. Much work still needs to be done in this
field particularly to improve our understanding of the relationship between
the structure and the photosensitivity of these photoinitiators. As the field
advances, one can expect still other new classes of sulfonium salt photoinitiators to be developed as well as continued improvements to be made
in the efficiency of the present systems. An understanding of the mechanism

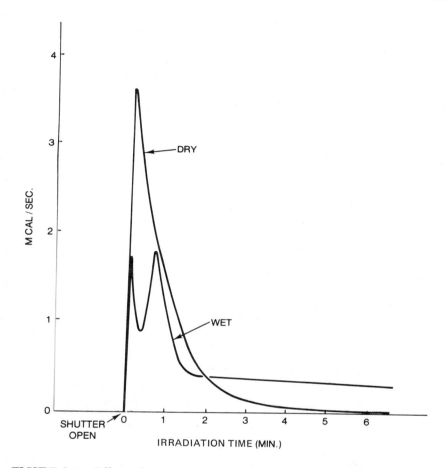

FIGURE 2.9 Effect of water on the photoinitiated polymerization of ERL 4221 containing 1.5 mol % $(C_6H_5)_3S^+ SbF_6^-$.

of photosensitization should lead to discovery of more efficient photo-sensitizers and a further broadening of their spectral response in photo-initiated cationic polymerizations.

Lastly, the process and polymers produced by sulfonium salt photo-initiated cationic polymerizations are eminently useful. Like the corre-sponding photoinitiated radical polymerizations which they complement, these cationic systems will find a wide range of applications where poly-merization speed and economy of energy utilization are of prime concern.

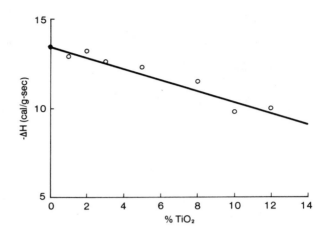

FIGURE 2.10   Effect of TiO$_2$ on the polymerization of ERL 4221 with 2%
$(C_6H_5)_3S^+$ SbF$_6^-$.

TABLE 2.3 [35]
Photosensitization of Onium Salts[a]

Photosensitizer[b]	E[c]	(M)[d]	Gel time, min Ph$_3$S$^+$ AsF$_6^-$
None	—	—	>85
Antracene	76	S	23
Benzophenone	69	R	>85
Perylene	66	S	24
Thioxanthone	66	T	>85
Phenothiazine	57	T	22

[a] 2×10^{-2} M (~1% by wt); 366 nm irradiation in 3,4-epoxy-
cyclohexylmethyl-3, 4'-epoxycyclohexane carboxylate:
acetone (2:1 by vol).

[b] 10^{-4} to 10^{-2} M (absorbance ~1).

[c] Excitation energies in kcal/mol.

[d] Spin multiplicity of reactive excited state.

REFERENCES

1. J. V. Crivello, in S. P. Pappas (ed.) UV Curing Science and Technology, Technology Marketing Corp., Stamford, Conn., 1978, p. 23.
2. J. J. Licari, W. Crepeau and P. C. Crepeau, U. S. Patent 3,205,157, September 7, 1965.
3. S. I. Schlessinger, U. S. Patent 3,708,296, January 2, 1973; 3,826,650, July 30, 1974.
4. W. R. Watt, U. S. Patent 3,794,576, February 26, 1974.
5. J. H. Feinberg, U. S. Patent 3,711,390, January 16, 1973; 3,816,281 June 11, 1974; 3,817,845, June 18, 1974 and 3,829,369, August 13, 1974.
6. S. I. Schlessinger, Photogr. Sci. and Eng., 18 (4), 384 (1974).
7. S. I. Schlessinger, Polym. Eng. and Sci., 14 (7), 513 (1974).
8. R. J. Ludwigsen, Soc. Mech. Eng., Technical Paper FC74-533 (1954); J. Rad. Curing, 2 (1), 10 (1975).
9. J. V. Crivello and J. H. W. Lam, J. Polymer Sci., Symp. No. 56, 1 (1976).
10. J. V. Crivello and J. H. W. Lam, Macromolecules, 10, 1307 (1977).
11. J. V. Crivello, U. S. Patent 3,981,897, September 21, 1976.
12. G. H. Smith, Belg. Patent 828,841, November 7, 1975.
13. Belg. Patent 837,782, June 22, 1976 to ICI.
14. J. V. Crivello and J. H. W. Lam, J. Polymer Sci., Chem. Ed., 17, 977 (1979). U. S. Patent, 4,058,401, November 15, 1977.
15. J. V. Crivello and J. H. W. Lam, J. Rad. Curing, 5 (1), 2 (1978).
16. G. H. Smith, U. S. Patent 4,069,054, January 17, 1978.
17. Belg. Patent 833,372, March 16, 1976 to ICI.
18. J. V. Crivello and J. H. W. Lam, J. Polymer Sci., Chem. Ed., 17, 1049 (1979).
19. J. V. Crivello and J. H. W. Lam, J. Polymer Sci., Chem. Ed., 17, 2877 (1979).
20. J. V. Crivello and J. H. W. Lam, J. Polymer Sci., Chem. Ed., 18, 1021 (1980).
21. A. Michaelis and E. Godcheux, Ber., 24, 757 (1891).
22. H. M. Pitt, U. S. Patent 2,807,648, September 24, 1957.
23. W. Hahn and R. Stroh, U. S. Patent 2,833,826, May 6, 1958.
24. K. Tsujihara, N. Furukawa, K. Oae and S. Oae, Bull. Chem. Soc. Jpn., 42, 2631 (1969).
25. P. Manya, A. Sekera and P. Rumpf, Bull. Soc. Chem., Fr., 1, 286 (1971).
26. D. S. Wildi, S. W. Taylor and H. A. Potratz, J. Am. Chem. Soc., 73, 1965 (1951).
27. G. H. Weigand and W. E. McEwen, J. Org. Chem., 33 (7), 2671 (1968).

28. C. B. May and Y. Tanaka, Epoxy Resin Technology, Marcel Dekker, Inc., New York, 1970, p. 199-205; F. W. Billmeyer, Jr., Textbook of Polymer Science, Interscience Pub., New York, 1964, p. 294.

29. J. W. Knapczyk and W. E. McEwen, J. Org. Chem., 35, 2539 (1970).

30. S. Torii, Y. Matsuyama, K. Kawasaki and K. Uneama, Bull. Chem. Soc. Jpn., 46, 2912 (1973).

31. J. W. Knapczyk, J. J. Lubinkowski and W. E. McEwen, Tetrahedron Letters, 3739 (1972).

32. J. E. Moore, S. H. Schroeter, A. R. Shultz, and L. D. Stang, A. C. S. Symp. Series No. 25, 9 (1976).

33. J. V. Crivello, J. H. W. Lam and C. N. Volante, A. C. S. Ctgs. and Plast. Preprints, 37 (2), 4 (1977).

34. J. V. Crivello, J. H. W. Lam and C. N. Volante, J. Rad. Curing, 4 (3), 2 (1977).

35. S. P. Pappas and J. H. Jilek, Photogr. Sci. and Eng., 23 (3), 141 (1979).

36. J. V. Crivello and J. H. W. Lam, J. Polymer Sci., Chem. Ed., 17, 1059 (1979).

THERMAL ANALYSIS OF THE CURING OF POLYESTER POWDER COATINGS WITH EPOXIES

R. van der Linde and E. G. Belder

Scado B. V.
Zwolle, The Netherlands

INTRODUCTION

The reaction of carboxylic acid groups containing resins with epoxy compounds constitutes an important curing mechanism in the coatings industry and in particular in modern powder coating technology. Polyester powder coatings cured with epoxies can be divided into two types, one mainly for interior and the other for exterior application. Apart from the polyester, these systems differ in the nature of the epoxy resin used for curing. For interior application Bisphenol-A-based epoxies are normally used; for exterior powder coatings Bisphenol-A-based epoxies are not suitable, because of UV-deterioration and therefore triglycidyl isocyanurate is used as the curing agent. The two systems are indicated in Figure 3.1.

Curing schedules in these systems were normally 10 min at 200°C. But in recent years the need for systems with lower temperature cure has become very strong, obviously due to higher energy prices. For that reason we have been working on the development of low cure polyester-epoxy powder coating systems of the above mentioned type. In this work we have used Differential Scanning Calorimetry (DSC) as an analytical tool.

55

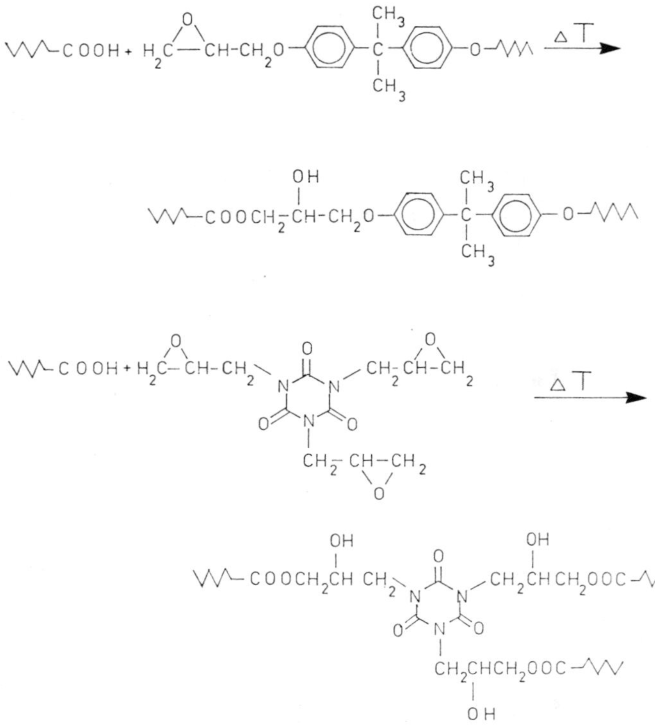

FIGURE 3.1   Polyester-epoxy powder coating for interior use and pure polyester–TGIC systems for exterior applications.

DIFFERENTIAL SCANNING CALORIMETRY

In quantitative Differential Scanning Calorimetry the heat flow of a sample is measured against a reference. A heat flow may result either from physical changes in the sample (e.g., melting, crystallization, glass-transition) or from chemical reactions. These changes can either be exothermic or endothermic. Figure 3.2 is a schematic representation of heat flow DSC.

There are two measuring modes—a dynamic mode, in which the sample is subjected to a linear temperature program, and an isothermal mode. Since it is possible to obtain kinetic data from a single dynamic run, dynamic DSC was favored for our measurements.

DSC-measurements were run on a Mettler TA 2000 apparatus. Samples were prepared by dissolving the polyesters and epoxy resins in a

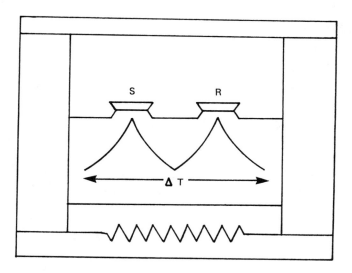

FIGURE 3.2   Schematic representation of measuring cell for Differential
Scanning Calorimetry.

stoichiometric ratio, adding the catalyst and evaporating the solvent.
Dichloromethane and acetone were used as solvents.

Figure 3.3 shows a typical dynamic run of a polyester-epoxy powder
coating formulation with a heating rate of 5°C/min. After a short equili-
bration period, we first observe the glass-transition point of the resin
mixture. For good physically stable systems this Tg should be >45°C.
Then at about 90°C there is another baseline shift, which has been attributed by Klaren [1] to melting of the powder to a continuous liquid (temperature of onset of flow, T_{of}) thus changing the contact between the material
and the pan. After that, depending on the reactivity of the system, an
exothermic peak is observed, starting at the temperature of onset of cure
(T_{oc}).

This exothermic curve provides the kinetic data for the reaction and
allows for the calculation of the following kinetic parameters: overall
reaction constant and its temperature dependence, activation energy, and
time for conversion at a defined temperature, using the following basic
equations:

$$k = \frac{dH}{dt} \frac{1}{H_r} \tag{1}$$

where $\frac{dH}{dt}$ = enthalpy change in mW, k = reaction constant in S^{-1}, and H_r =
remaining peak area (Figure 3.3) in mJ.

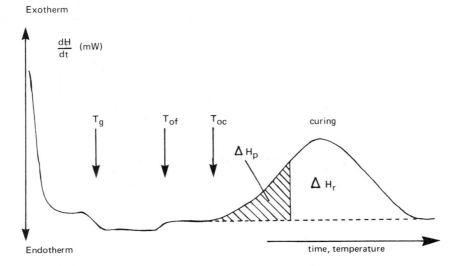

FIGURE 3.3   Typical DSC-curve of a polyester-epoxy powder coating
composition (dT/dt = 5°K/min).

$$k = k_o \exp \left[\frac{-E_A}{RT} \right] \qquad (2)$$

where k_o = frequency factor in s^{-1}, E_A = activation energy in kJ/mol,
R = gas constant 8.31 J/mol °K, and T = absolute temperature.

$$\frac{d\alpha}{dt} = k (1-\alpha)^n \qquad (3)$$

where $\frac{d\alpha}{dt}$ = reaction rate in S^{-1}, α = reaction conversion (from 0 to 1),
and n = order of the reaction.

KINETIC MODELS FOR THE CARBOXYLIC
ACID-EPOXY REACTION

Before going into the results of the thermal analysis we will discuss
the underlying kinetics of the carboxylic acid-epoxy reactions (see also
[2,3]). Apart from the uncatalysed and the acid catalysed acid-epoxy reaction, which we will not be dealing with in this paper, there are two
fundamentally different reaction mechanisms for the base catalysed acid-
epoxy reaction.

Prior Equilibrium Followed by a "Slow" Reaction

$$B(ase) + RCOOH \underset{}{\overset{K_{eq}}{\rightleftharpoons}} BH^{+} + RCOO^{-} \tag{4}$$

$$RCOO^{-} + H_2C\overset{O}{\underset{}{\diagdown}}CH- \overset{k_1}{\longrightarrow} RCOOCH_2\overset{O^{-}}{\underset{}{\diagup}}CH- \tag{5}$$

$$RCOOCH_2 - \overset{\overset{O^{-}}{|}}{C}H- + BH^{+} \overset{fast}{\longrightarrow} RCOOCH_2 - \overset{\overset{OH}{|}}{C}H- + B \tag{6a}$$

and/or

$$RCOOCH_2 - \overset{\overset{O^{-}}{|}}{C}H- + RCOOH \overset{fast}{\longrightarrow}$$

$$RCOOCH_2 - \overset{\overset{OH}{|}}{C}H- + RCOO^{-} \tag{6b}$$

Since both (6a) and (6b) may be expected to be faster than (5), the rate expression for this mechanism is

$$S_2 = k_1 [RCOO^{-}][epoxy]$$

$$= k_1 K_{eq} \frac{[RCOOH][epoxy][B]}{[BH^{+}]} \tag{7}$$

This is an example of specific base catalysis, because the rate will be dependent on the equilibrium constant in equation (4) and therefore be specific as regards the type of base. This mechanism is thought to be valid for catalysis by Lewis bases such as tertiary amines.

Direct Attack of the Base on the Epoxy

$$B^{-} + H_2C\overset{O}{\overset{\diagup\diagdown}{}}CH- \overset{k_2}{\longrightarrow} BCH_2 - \overset{O^{-}}{\overset{\diagup}{}}CH- \tag{8}$$

$$BCH_2 - \overset{\overset{O^{-}}{|}}{C}H- + RCOOH \overset{fast}{\longrightarrow} BCH_2 - \overset{\overset{OH}{|}}{C}H- + RCOO^{-} \tag{9}$$

$$RC OO^- + H_2C - CH- \xrightarrow{k_3} RC OOCH_2 - CH - \qquad (10)$$

$$RC OOCH_2 - CH- + RC OOH \xrightarrow{fast} RC OOCH_2 - CH- + RC OO^- \qquad (11)$$

Prior to (7) a dissociation step might be involved e.g.,

$$A^+B^- \xrightleftharpoons{K_{eq}} A^+ + B^-$$

but since B^- is reacting irreversibly with epoxy this equilibrium will shift completely to the right and does not play a role in the kinetics. The rate expressions for this mechanism are

$$S_2 = k_2 [B^-] [epoxy]$$

$$S_3 = k_3 [RC OO^-] [epoxy]$$

Both for $k_2 < k_3$ and $k_2 > k_3$, $[RC OO^-] = [B^-]_{original}$ = constant, because B^- is disappearing (8) and $RC OO^-$ is regenerated (11), and therefore $S_3 = k_3' [epoxy]$.

When $k_2 < k_3$ there is a delay before the reaction runs at its expected rate. We believe that this mechanism applies for catalysis of the acid-epoxy reaction with e.g., halides. This is supported by the results obtained in solution from the reaction of benzoic acid and phenyl glycidyl ether [4], catalyzed by a variety of halide catalysts, where it was found (1) the rate is not influenced by the nature of the halide. Cl^-, Br^- and I^- gave almost the same k-value, and (2) at 132°C an "incubation" time of about 1 h was observed, before the reaction displayed perfect first order kinetics, as shown in Figure 3.4.

Obviously the above described mechanisms are simplified and do not acknowledge the occurrence of side reactions, which certainly exist. One side reaction, which becomes more important with the extent to which the acid-epoxy reaction proceeds, is etherification according to (12)

$$RC OOCH_2 - CH - + H_2C - CH \longrightarrow RC OOCH_2 - CH \qquad (12)$$

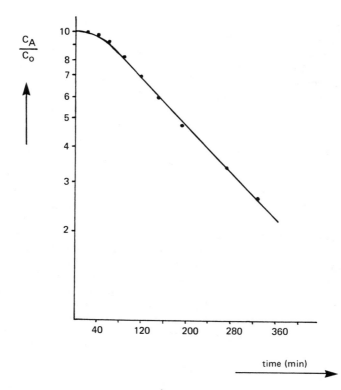

FIGURE 3.4   Kinetic plot of $\ln C_A/C_o$ against time for the reaction of
benzoic acid (C_A) with phenylglycidyl ether at $132\,^\circ C$ in
$CH_3OCH_2CH_2OCH_3$.

Moreover, the nature of A^+ in the second mechanism might influence the
scheme via (13)

$$RCOO^- + A^+ \rightleftharpoons RCOO^-A^+ \tag{13}$$

thus reducing the actual $[RCOO^-]$ for reaction, or via (14),

$$RCOO^- + A^+ \xrightarrow{k_4} RCOOH + A \tag{14}$$

when A^+ has an abstractable proton. In this case, depending on k_4, the
reaction is self-extinguishing.

When considering powder coating compositions, application of the
above kinetics is certainly not allowed after the reaction has proceeded to

a large extent and the viscosity of the system has become high. At that
stage the reaction rate becomes diffusion controlled.

Differential Scanning Calorimetry provides the overall reaction
constant, without differentiating between the various mechanisms. This
reaction constant can be regarded as an average reaction constant over the
whole reaction range. It is often referred to as the applied reaction
constant.

DSC RESULTS OF SOME BASE CATALYZED POLYESTER-EPOXY
POWDER COATING COMPOSITIONS

From the curve shown in Figure 3.3, one obtains the overall reaction
constant k and its temperature dependence from which an Arrhenius plot
can be constructed. This is shown in Figure 3.5 for a polyester-epoxy
resin composition catalyzed by 0.8% and 1.2% dimethyl benzylamine. The
important feature of this plot is the value for β, which is directly proportional to the activation energy of the reaction. Going from 0.8% to 1.2%
catalyst the overall reaction constant increases as expected, but is more
pronounced at low temperature than at high temperature, thus giving rise
to a lower activation energy of 60.1 kJ/mol compared to 65.4 kJ/mol for
0.8% catalyst. The effect on the reaction rate can easily be derived from
this plot. The 1.2% catalyzed system reacts as fast at e.g., 153°C as the
0.8% catalyzed system at 160°C.

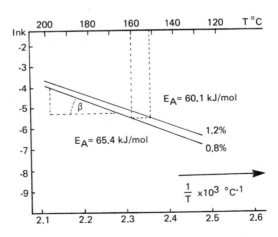

FIGURE 3.5   Arrhenius plots for two polyester-epoxy powder coating
compositions catalyzed by 0.8 and 1.2% (W/W on polyester)
dimethyl benzylamine.

As a second example we prepared a series of polyester-epoxy powder coating compositions containing an increasing level of tetramethyl ammonium chloride from 0.12 to 0.24% (based on polyester weight). The Arrhenius plots of these compositions are shown in Figure 3.6. One obtains a set of four almost parallel lines, from which the reaction constant at e.g., 130°C can be evaluated. The results are contained in Table 3.1 together with the activation energy. The values found for the activation energy are, as might be expected, somewhat lower than those found by Henig et al. [5] for the uncatalyzed polyester-epoxy cured reaction.

If we plot k^{130° against the catalyst concentration a linear relationship is indicated, as expected from the reaction mechanism discussed before [equations (7) to (10)], where

$$k_3' = k_3 \,[\text{cat}]$$

This is shown in Figure 3.7.

Finally, we would like to present the results obtained with some recently developed products. The objective was to develop a range of products having curing schedules ranging from 20 min at 200°C for extremely good flow to 15 min at 140°C for low temperature cure. The DSC curves of these products are shown in Figure 3.8.

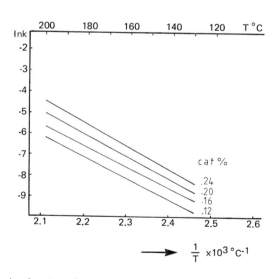

FIGURE 3.6   Arrhenius plots for four polyester-epoxy powder coating compositions with increasing level of tetra methylammonium chloride as a catalyst.

TABLE 3.1

% cat	$k^{130°} \times 10^4 \ s^{-1}$	E_A kJ/mol
0.12	0.52	84.8
0.16	1.14	81.4
0.20	1.58	80.0
0.24	2.25	88.9

Figure 3.9 shows time-temperature plots for each of these products at a reaction conversion of 90%. From these plots one can easily derive the optimum curing temperature at any desired curing time. So product A may be cured at 20'/140°C of 10'/160°C. Product B will give good results at 20'/145°C or 10'/165°C and so on.

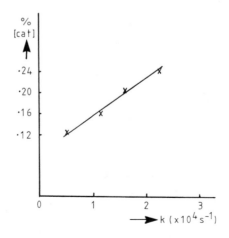

FIGURE 3.7   Plot of k (130°C) against catalyst concentration showing
linear relationship.

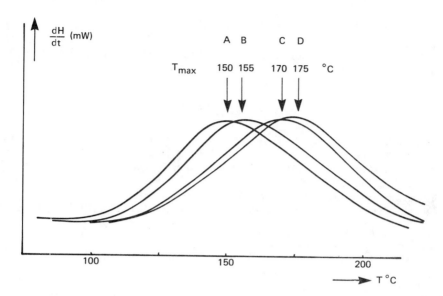

FIGURE 3.8   DSC-curves for four commercial polyester-epoxy compositions.

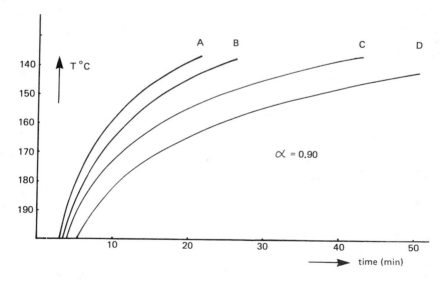

FIGURE 3.9   Plot of reaction conversion against time for four commercial polyester-epoxy compositions.

ACKNOWLEDGMENT

The authors wish to thank Mr. A. Westendorp for his assistance with the experimental work.

REFERENCES

1. C. H. J. Klaren, J. Oil. Colour Chem. Assoc., 60, 205 (1977).
2. L. Shechter and J. Wijnstra, Ind. Eng. Chem., 48, 86 (1956).
3. C. A. May and J. Tanaka, Epoxy Resins, Marcel Dekker Inc., New York, 1973, p. 178.
4. A. Visser and R. O. de Jongh, unpublished results.
5. A. Henig, J. Jäth and H. Möhler, Farbe und Lack, 86, 313 (1980).

RHEOLOGY OF EPOXY RESIN-BASED POWDER FILMS

S. A. Stachowiak

Koninklijke/Shell-Laboratorium, Amsterdam
Shell Research B. V.
Amsterdam, The Netherlands

INTRODUCTION

Powder coating is a versatile method of applying protective and decorative industrial films. Currently, powder coatings are employed in a spectrum of applications, and this growing acceptance is reflected in the steadily increasing share taken up by powders in industrial coatings. It is very likely that this trend will continue in the future, at least partly because of the stricter environmental legislation currently being adopted by pollution-conscious authorities. In view of Shell's position as a major manufacturer of epoxy resin, it was decided that a study should be made of various aspects of the preparation and application of epoxy-based powder coatings to ensure that the quality of the EPIKOTE resins available to our customers remains as high as possible.

One of the aspects we have studied is the appearance of powder films, i. e., the degree of flow, the occurrence of cratering, and other film defects. This has, of course, been the subject of a number of studies reported in the literature. However, we have concentrated on epoxy-based powders and this paper considers the effects of so-called "flow-promoting" additives on the melt viscosity and surface tension, and hence on the appearance, of powder films.

FILM FORMATION AND FLOW DURING STOVING

Powder coatings are unique in the field of industrial paints because of the phase changes which occur during both powder manufacture and film formation. The rheological forces which come into play during these phase changes determine to a large extent the final appearance of the film.

During extrusion it is important that the pigment agglomerates present are efficiently broken up and held apart until the dispersed structure is "frozen in" on cooling. On stoving, the powder particles must first fuse together as melting of the resin component takes place [1]. This is shown in Figure 4.1.

Several theoretical approaches to coalescence have been developed [2], although it is rather difficult to apply these theories exactly to normal powders, since these consist of irregularly-shaped particles of varying size. It is clear, however, that the most important factors governing the speed of coalescence (at a given temperature) are (a) melting point of the resin; (b) viscosity of the molten powder particles; and (c) radius of curvature of the molten powder particles. For optimal flow the coalescence should take place as quickly as possible so that as much time as possible is available for the leveling-out stage, illustrated in Figure 4.2.

It has been shown [3] that the critical factors influencing the leveling-out of a newly-formed film are surface tension, melt viscosity, and film thickness. The melt viscosity is partly determined by variables such as the stoving temperature, the cure speed, and the heating-up rate. These factors, together with the particle size distribution and the film thickness, are in turn usually determined by the required mechanical properties of the film, the nature of the object being coated, the manufacturing and application equipment available, etc. Thus, from the point of view of the epoxy-resin manufacturer, it is important to make a resin which will provide a molten film having the appropriate surface tension and melt viscosity to ensure that leveling-out is optimal under all operating conditions.

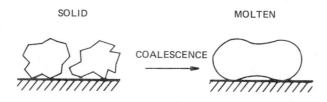

FIGURE 4.1   Schematic diagram showing the coalescence of molten powder particles.

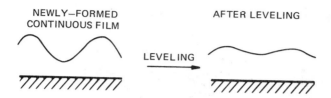

FIGURE 4.2   Schematic diagram showing the leveling-out of a newly-
            formed powder film.

FACTORS AFFECTING FILM FLOW AND APPEARANCE

The driving force for flowout in powder coatings is undoubtedly the
surface tension of the system, which attempts to straighten out the sinu-
soidal curve shown in Figure 4.2 and thus reduce the overall surface area.
This force is resisted by the intermolecular attractions exerted in the
film, so that the higher the melt viscosity, the more resistance will be
offered to flowout. Thus, the magnitude of the difference between the
forces arising from the surface tension and the intermolecular attractions
will determine the extent of flowout.

In order to maximize this difference the surface tension should
obviously be as high as possible, and the melt viscosity as low as possible.
This could be achieved by incorporating an additive which would increase
the surface tension of the system, and using a resin having a lower molecular weight and melting point. Such a combination would also serve to
reduce the time required for coalescence to take place. However, it is
unlikely that this formulation would be successful; although the coating
would exhibit an excellent flow, this would inevitably be accompanied by
cratering, due to the high surface tension, as well as sagging and poor
edge coverage due to the excessively low melt viscosity.

In practice, we are essentially confined to the surface tension and
melt viscosity limits illustrated in Figure 4.3. Excessively low surface
tension and/or high melt viscosity will clearly preclude efficient film
flowout.

Let us consider the restraints placed on these properties in the
opposite directions. The upper limit of allowable surface tension is
delineated by the onset of cratering during film formation. Cratering in
paint films (though not in powder coatings) has already been studied [4-6].
The subject is complex but it is clear that a proposed mechanism [4] of
liquid-liquid spreading at the paint film/air interface forms a likely explanation of this phenomenon. It is suggested that the presence of traces of an
immiscible liquid on the paint surface can lead to cratering if the surface
tension of the paint film (γ_P) is greater than the sum of the surface tension
of the immiscible liquid ($\gamma_{\bar{L}}$) and the interfacial tension between the paint
film and the immiscible liquid ($\gamma_{\underline{LP}}$), as illustrated in Figure 4.4.

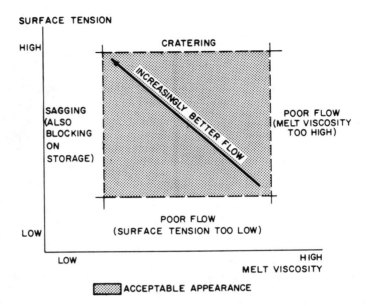

FIGURE 4.3   Dependence of film-flow on magnitudes of surface tension
and melt viscosity.

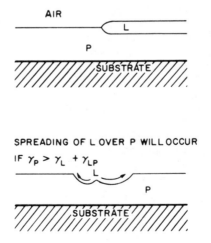

FIGURE 4.4   The spreading-out of an immiscible impurity over a molten
powder film.

This rapid spreading is thought to result in the bulk transfer of under-lying paint along with the spreading immiscible liquid. In the case of a thin film this cannot be compensated for by return flow and so a crater is formed. With thicker films some return flow is possible and cratering is therefore not such a problem. This phenomenon of bulk transfer as a re-sult of local differences in surface tension is known as the Marangoni effect and is also the cause of other paint defects such as Bénard cells in solvent-based paints.

The treatment for cratering in all cases in the references above was the same and was immediately effective: a lowering of the surface tension of the paint film. This ensures that γ_P is always less than $\underline{\gamma_L + \gamma_{LP}}$, so that the flowing-out of any immiscible impurities on the surface of the paint film does not take place.

Low-surface tension impurities on the substrate can also form a source of cratering by either migrating to the paint film/air interface or remaining as a low surface tension area on the substrate which cannot be wet by the molten powder coating. Studies by Zisman and others [7-10] have shown that for a liquid to wet a substrate its surface tension must be lower than the critical surface tension of the substrate. The treatment for this type of cratering is the same as above, viz a lowering of the surface tension of the paint film. The presence of impurities on a steel panel after solvent degreasing was demonstrated by Johnson [11], who showed the presence of a phthalate ester, probably a mill-processing aid, on such a panel.

The effect of surface tension on cratering is shown in Figure 4.5. Here, resins of increasing molecular weight (and therefore increasing surface tension) in the order EPIKOTE 1002, EPIKOTE 1055, EPIKOTE 1007 have been powdered, sieved, and applied by electrostatic spraying to steel panels which had been degreased together in the same batch. It can be seen that the severity of cratering increases with increasing molecular weight (and therefore increasing surface tension) of the resin. In the case of EPIKOTE 1055, however, reducing the surface tension by addition of an anticratering agent eliminates the cratering completely. We have studied the effectiveness of various anticratering agents and the accompanying effect on film flow. This is discussed below.

It is a simple matter to improve film flow by using a base resin of lower molecular weight. However, if the melt viscosity is too low, the physical storage stability of the powder will be poor and, on application, poor edge coverage and sagging on vertical surfaces may result. A low melt viscosity coupled with good physical storage stability may however be achieved by improving the degree of pigment dispersion within the powder. It has been shown [12] that the viscosity of a given dispersion of a pigment in a medium is dependent on how finely the pigment is dispersed within the medium. In the case of a medium which wets the pigment only poorly, the fineness of the dispersion can be improved dramatically—and hence the melt viscosity lowered—by addition of a dispersion aid.

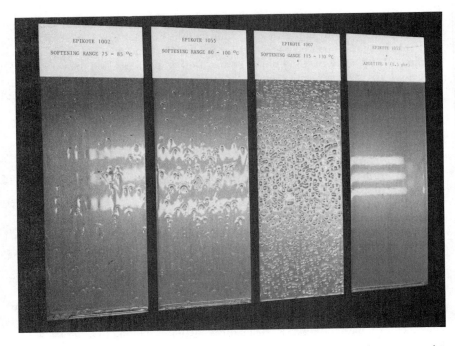

FIGURE 4.5   Formation of craters in unpigmented, uncured epoxy-resin films.

During dispersion the molecules of the dispersion aid are attracted to the newly-formed pigment surfaces, where they then present a surface which is more receptive to the wetting action of the medium. In powder coatings this means that, if necessary, a higher-melting base resin may be used, since once the resin has melted during stoving, the viscosity of the liquid film is low owing to the high degree of pigment dispersion, and good flow will be obtained. We have studied the use of such dispersion aids in powder coatings and these findings are discussed below.

EFFECTIVENESS OF ANTICRATERING AGENTS AND THEIR EFFECT ON FILM FLOW

In its simplest form, a powder coating consists of base resin, pigment, and curing agent. This basic combination will yield the type of film shown in Figure 4.6. As discussed earlier, the appearance of a film can be improved by the addition of an anticratering agent which, by decreasing the surface tension, eliminates cratering and reduces poor-edge coverage.

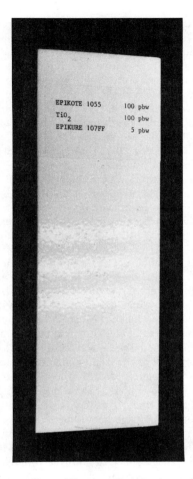

FIGURE 4.6   Powder coating without anticratering agent.

Unfortunately, this reduction in surface tension also means that the driving force for flowout is decreased, which can result in poorer flow.

In order to study this more closely, a series of powders was prepared in which only the nature of the anticratering agent was varied. The following standard powder formulation was chosen:

	pbw
EPIKOTE 1055	100
TiO$_2$ (Kronos C L 220)	100
EPIKURE 147	4.5
Anticratering agent	0.5

The relatively high pigment level was chosen in order to highlight differences in flow caused by the presence of the various anticratering agents. EPIKURE 147 was selected as the curing agent because it is a low-melting, single substance and should therefore not give rise to extraneous flow or cratering effects due to incomplete dispersion during extrusion. A number of commercially available acrylate type anticratering agents were incorporated into this series of powders. The powders were extruded through a Buss Ko Kneader PLK 46 extruder, ground, sieved through a 100 μm sieve, and applied using an electrostatic spray to 0.5 mm degreased, cold-rolled steel panels. After stoving for 15 min at 180°C, and 50-60 μm thick films were assessed for cratering and extent of flow.

The cratering assessment was straightforward: no cratering was observed in any film. The assessment of the extent of flow, or alternatively, the degree of orange peel, was done visually by comparison with a set of standard reference panels exhibiting flows rated from mirror-like (1) through good (4) to poor (8-10). Some of these standard panels are shown in Figure 4.7.

The anticratering agents examined are shown in Table 4.1. Their molecular weights are spread over a wide range and at least three of them are based on exactly the same monomers. We had already studied [13] additives A, B, and C in powders based on EPIKOTE 1004 but having a pigment level of only 50 phr. In that formulation, equivalent flow was obtained with all three additives. The flow obtained with the present formulation is shown in Table 4.2. Here considerable differences in flow were observed, probably because of the higher pigment loading. A clear trend of better flow with decreasing molecular weight of the additives was found. This type of behavior had been noted previously [14], although in a different system. It would seem that, if excessive flow deterioration is to be avoided, the anticratering additive should be present at the film surface at the minimum concentration required to prevent cratering. This suggests that for an optimal effectiveness, the additive should have a borderline compatibility with the base resin. An additive which, at a given concentration, is too soluble in the base resin will not exude to the surface and thus will have little effect on the surface tension of the molten system. On the other hand, an additive with too little solubility in the resin will exude completely to the surface, thereby dramatically reducing the surface tension and hence the flow. Addition of larger amounts of the too soluble type may be expensive and result in the deterioration of other properties, while addition of smaller amounts of the incompatible type may be awkward to carry out on an industrial scale, or might even lead to cratering. Of the additives tested, additive E therefore seems to display the best combination of structure and molecular weight at the concentration applied.

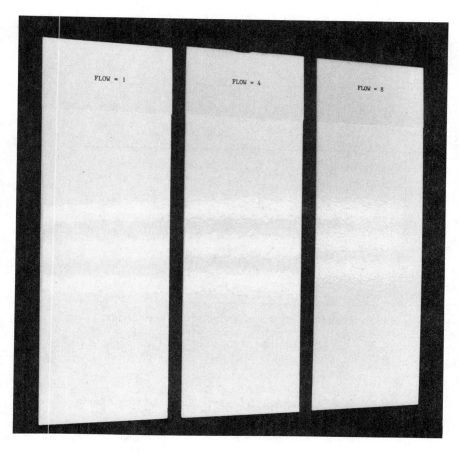

FIGURE 4.7   Films exhibiting excellent (1), good (4), and poor (8) flow.

EFFECTIVENESS OF FLOW-PROMOTING ADDITIVES

In the introduction, we described how the addition of a dispersion aid could improve the degree of pigment dispersion and hence the flow. We decided to investigate the effectiveness of such flow-promoting additives in powder coatings by incorporating such aids into the standard powder-coating formulation used for the evaluation of the anticratering agents.

Two flow-promoting additives were evaluated, a fluorinated ester (product F) and a product described by its supplier as a reactive additive for epoxy resins (product G). The results of the evaluation are given in Table 4.3. An improvement in flow was indeed observed on addition of the

TABLE 4.1
Liquid Acrylate-Type Anticratering Agents
Used in Powder Coatings

Anticratering agent	Structure	Mn $(\times 10^3)$	Mw* $(\times 10^3)$	Mz $(\times 10^4)$
A	n-Butyl acrylate homopolymer	17.5	116.9	99.1
B	Ethyl acrylate/	16.9	54.8	64.5
C	ethyl hexyl acrylate	7.8	29.2	15.5
D	copolymer	5.3	12.0	3.5
E	n-Butyl acrylate/ ethyl hexyl acrylate copolymer	1.5	4.9	4.1

*Molecular weights from GPC with polystyrene calibration.

TABLE 4.2
Effect of Liquid Acrylate-Type Anticratering Agents
on Flow in Powder Coatings[a]

Anticratering agent	Mn $(\times 10^3)$	Flow[b]
A	17.5	9
B	16.9	9
C	7.8	7
D	5.3	7
E	1.5	5

[a]In all cases concentration of additive = 0.5 phr.

[b]Flow assessment: 1 = excellent; 4-5 = good; 8 = poor.

TABLE 4.3
Effect of Flow-Promoting Additives
on Flow in Powder Coatings

Additive	Type	Concentration (pbw)	Flow
B F G	Acrylate Fluorinated ester Reactive additive for epoxy resins	0.5 0.3 1.0	9 5 4

Powder formulation (pbw):   Flow assessment:
  EPIKOTE 1055   100   1 = excellent
  TiO2   100   4-5 = good
  EPIKURE 147   4.5   8 = poor
  Additive   0.3-1.0

flow-promoting agents. Some cratering was observed in the film containing product G. It was decided to investigate this flow improvement a little more deeply.

Types of Flow

At this point, it is useful to consider the types of viscosity behavior which can occur. The simplest type of behavior is found with materials exhibiting Newtonian flow. In this case, a plot of shear rate against shear stress is found to be a straight line passing through the origin so that at constant temperature, the viscosity, which is defined as shear stress divided by shear rate, is found to be practically constant at low- and medium-shear rates. It can be represented by the simple relationship, viscosity = $\tau/\dot{\gamma}$, typically expressed in Pa.s. Other materials, however, do not exhibit this straightforward behavior. In these cases, our simple equation is not sufficient to describe the viscosity at a given temperature under varying shear rates.

Plastic flow is an example of such anomalous flow behavior. Here, a plot of shear rate against shear stress is approximately a straight line, but one which does not pass through the origin, as shown in Figure 4.8. This type of flow is very important for us because it is the type of flow

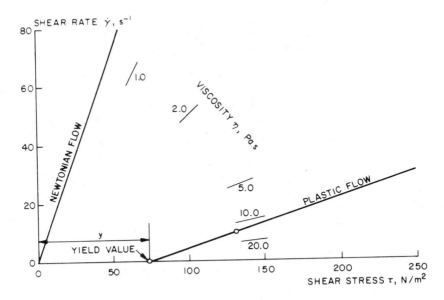

FIGURE 4.8   Examples of Newtonian and plastic flow.

often exhibited by dispersions of TiO_2 in resins.   Plastic flow introduces
the idea of a yield value, i.e., a certain minimum shear stress which
must be exceeded before flow will take place.   This minimum shear stress
is given by the intercept on the shear stress axis (y).   The presence of a
yield value indicates the existence of considerable interparticle forces.   In
the case of a dispersion of titanium dioxide in the resin, attraction between
pigment particles is undoubtedly responsible for the build-up of these forces
and it is clear that the magnitude of these forces will be determined by the
efficiency with which the pigment particles have been wet.   In the case of a
molten powder coating one can imagine the initial surface irregularities
being smoothed out under the influence of the surface tension, but that as
the driving force for flowout decreases, the presence of a significant yield
value could prevent further flowout, resulting in the "freezing-in" of the
incompletely flowed-out structure.

The plastic viscosity (μ) can be obtained from the slope of the
shear rate/shear stress line in Figure 4.8:

$$\mu = \frac{(\tau - y)}{\dot{\gamma}} \qquad \mu = \frac{\tau}{\dot{\gamma}} - \frac{y}{\dot{\gamma}} \tag{1}$$

giving us a relationship between absolute viscosity and plastic viscosity:

$$\eta = \mu + y/\dot{\gamma} \ .$$

(2)

The Weissenberg Rheogoniometer

In the case of systems exhibiting plastic flow, we are interested in
determining the yield value and plastic viscosity. It is evident from
Figure 4.8 that in order to be able to do this accurately, shear stresses
must be measured at very low shear rates. This is doubly important in
the case of powder coatings since it is under these exceedingly low shear
rates that flowout takes place. Several commercial viscometers are
available, among them the Weissenberg Rheogoniometer (WRG). This
instrument was chosen for our studies since it can be used to measure
shear stresses at these very low shear rates ($10^{-3} - 10^{-2}$ s^{-1}). A limitation, however, is that only uncured powders may be quantitatively studied
in the WRG, since during testing the sample must remain at an elevated
temperature for a considerable time.

Figure 4.9 shows a schematic diagram of the cone and plate assembly
of the WRG. The sample is placed between the cone and plate, which are
housed in an oven. Half an hour is usually allowed for the sample to reach
thermal equilibrium at 150°C. A known speed of rotation is applied to the
cone which results in the development of shear within the sample. The

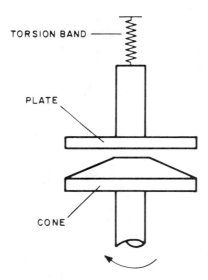

FIGURE 4.9   Cone and plate arrangement of the WRG.

shear stress can be calculated by measuri ng the relative motions of cone
and plate. This process is repeated for a series of speeds of rotation of
the cone so that a plot of shear rate against corresponding shear stress
can be drawn. A typical example of such a run is shown in Figure 4.10.
This shows that three values of shear stress may be observed, the maxi-
mum value τ_M, the equilibrium value τ_∞ and the residual tension τ_R. This
is in agreement with the idea that within thepigment dispersion there exists
a network of agglomerated pigment particles which is broken up under the
applied shearing force (τ_M is the apparent force required to do this). Once
the interparticle forces are broken and the particles are in motion, a
smaller force (τ_∞) is required to maintain the flow. On stopping the rotation, particle-to-particle contacts can be renewed; these prevent the plate
from relaxing completely back to the original position and are responsible
for the residual tension τ_R.

Every powder sample should therefore yield three values of τ for
each shear rate applied. However, these τ values are only valid if they
have been determined at a constant shear rate. Now, while this is the
case with τ_∞ and τ_R, it does not hold for τ_M, since the very small rotation of the plate relative to the cone as the τ_M is reached and passed is
sufficient to cause the actual shear rate to deviate considerably from the
shear rate as calculated from the speed of the cone.

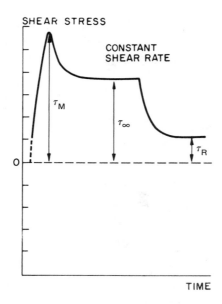

FIGURE 4.10   Shear stress as a function of time at a fixed shear rate as
measured using the WRG.

The conversion of the valid shear stress measurements into yield value and plastic viscosity is illustrated in Figure 4.11. Here the τ_∞ (open circles) and τ_R (closed circles) values obtained for a given powder coating are plotted as a function of the shear rate ($\dot{\gamma}$). In view of the possible correlation between τ as $\dot{\gamma}$ approaches zero and the extent of flowout, the curves have been extrapolated to zero shear rate, it being borne in mind that the extrapolations for τ_∞ and τ_R should both lead to the same yield value. Once the yield value has been fixed it is then possible to calculate the plastic viscosity by measurement of the angle β. As will be seen later, this angle β can be difficult to assess and so most use has been made of yield value comparisons.

Powder Formulations Studied

The influence on the powder rheology of the pigment level, the presence of anticratering agent and the presence of flow-promoting agent were studied simultaneously by measuring the yield values and plastic viscosities of the following powders:

- EPIKOTE 1055
- EPIKOTE 1055/Product B (0.5 phr)   with 50, 100 and 140 phr
- EPIKOTE 1055/Product G (4.0 phr) TiO_2

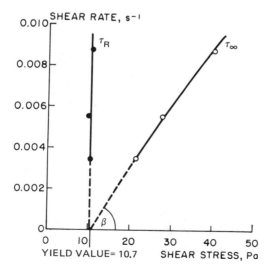

FIGURE 4.11   Derivation of yield values and plastic viscosities from τ_∞ and τ_R measurements.

The yield values and plastic viscosities were calculated as shown in Figure
4.11. It is interesting to study these yield values by plotting the logarithms
of the yield values against pigment concentration, as shown in Figure 4.12.

First of all let us consider the powders containing only resin and pigment, and those with added anticratering agent (product B). In both cases
it is quite clear that the yield value increases sharply with increasing pigment level. The presence of product B does not appear to have a significant effect on the yield values of these powders. It is found, however, that
the flow deteriorates on addition of product B. This is attributed to the
accompanying reduction in surface tension which eliminates cratering but
simultaneously reduces the driving force for flowout.

The addition of flow-promoting agent (product G), on the other hand,
has a marked influence on the yield value at lower pigment concentrations.
This correlates well with the flow obtained with these powders. For example, the powder containing 50 phr pigment exhibited an especially good
flow. This is attributed to an improvement in the degree of pigment dispersion within the powder. With this powder no τ_M was observed in the
shear stress traces, which indicates a large reduction in the number of
interparticle contacts.

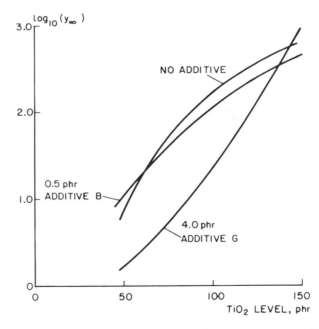

FIGURE 4.12   Log$_{10}$(yield value) versus pigment level for powders with
and without additive.

Figure 4.13 shows the shear rate/shear stress diagrams for the powders with 100 phr pigment. It can be seen that the τ_∞ line increasingly deviates from a straight line as the yield value increases. This deviation is further accentuated in the powders with 140 phr pigment. Because of these deviations it is not always possible to determine the plastic viscosity of a system by measuring the angle β. However, the plastic viscosity of each system can be estimated by measuring the slope at a given shear rate, say, 0.002 s^{-1}; it is then found that the same general trend is followed as with the yield values.

We have seen how the pigment concentration and the presence of a flow-promoting agent can affect the rheology of a powder. The powders were also applied to steel panels and stoved at 150°C for 15 min. The films were then visually assessed for extent of flow and cratering. Powders containing product $\underline{F}$ and a lower level of product $\underline{G}$ were also included. The films containing products $\underline{B}$ and $\underline{F}$ were crater-free. Those containing

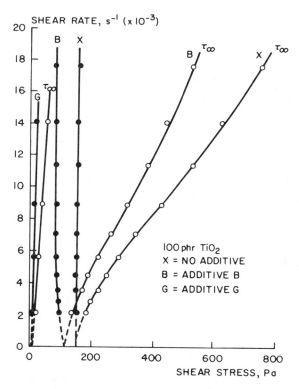

FIGURE 4.13   Shear rate/shear stress diagrams for powders containing 100 phr pigment.

TABLE 4.4

Correlation Between Yield Value and Flow of Uncured
Powder Coatings Based on "EPIKOTE" 1055

| Pigment concentration (phr) | Additive | | Yield value (Pa) | Flow[a] |
	Type	Concentration (phr)		
50	G	4.0	2	1
	None	—	7	7
	B	0.5	11	7
100	G	4.0	23	8
	G	2.0	97	9
	F	0.3	97	9
	None	—	167	10
	B	0.5	113	11
140	G	4.0	413	11
	None	—	520	11
	B	0.5	373	12

[a]Flow assessment: 1 = excellent; 4-5 = good; 8 = poor.

no additive or product G showed a tendency to cratering. Table 4.4 shows
the ranking given for the observed flow, together with the corresponding
yield values. We were particularly interested to see whether a correlation existed between yield value and extent of flow.

Correlation Between Yield Values and Observed Flow

In Table 4.4, the powders are arranged in order of deteriorating
flow. In practically every case it can be seen that increasing yield value
results in poorer flow. The two exceptions must be attributed to the surface tension-lowering effect of acrylate additive B. We have also examined
other powder systems in this way and can conclude that the flow can be
predicted from the yield value and that four main areas of yield value can
be distinguished: 0-3 Pa, very good to excellent flow; 3-7 Pa, moderate
flow; 7-50 Pa, poor flow; >50 Pa, very poor flow. These divisions are
not rigorous, of course, and will vary according to the magnitude of the

surface tension; however, they do give an indication of the barrier to flow
formed by the yield value which must be overcome before flowout can
occur.

Effect of Curing Agent

        We now consider the influence exerted by the curing agent in deter-
mining the flowout of a powder coating. We have seen how the degree of
pigment dispersion is important in determining the extent of flow. The
presence of curing agent reduces the time available for flowout to take
place; it is well known that exceptionally reactive powders often exhibit
poor film flow. However, the curing agent can also dramatically affect
the degree of pigment dispersion.
        Table 4.5 shows that the addition of 5 phr EPIKURE 107FF to the
powders with 50 to 100 phr pigment in almost all cases results in a remarkable improvement in flow. It can be imagined that the curing agent
also acts partly as a pigment-wetting promoter, improving the degree of
pigment dispersion during the extrusion step and thereby decreasing the
melt viscosity during stoving. In the case of the powder exhibiting excellent
flow, the pigment dispersion is already very good, so that addition of curing
agent only serves to increase the viscosity via crosslinking.
        We looked briefly at the rheological properties of these cured powders
using the WRG. Owing to the presence of the curing agent, only one run
could be carried out with each sample. Therefore, a different experimental procedure was adopted: measurements were begun approximately
1 min after the sample had been placed in the oven and the development of
shear stress at one particular shear rate, 0.221 s^{-1}, at a temperature of
150°C was studied. The logarithm of the viscosity was plotted against
time and in most cases straight lines were obtained which could be extrapolated back to give the viscosity at zero time. The values obtained are
given in Table 4.6, and the effect of curing agent on the flow can be seen
in Figure 4.14.
        The viscosity-lowering effect of the curing agent is obvious. This is
not too surprising in view of the amino moieties present in an accelerated
dicyandiamide-type curing agent such as EPIKURE 107FF. The efficiency
of this pigment-wetting effect will, of course, vary from curing agent to
curing agent, so that when a powder coating is to be formulated, it is
important that this flow-enhancing ability of the curing agent be taken into
account along with its reactivity.

Conclusions

        The following conclusions may be drawn from the work with uncured
powders described in this section:

TABLE 4.5
Flow Obtained with Powder Coatings Based on "EPIKOTE" 1055
With and Without Curing Agent

Pigment concentration (phr)	Additive Type	Concentration (phr)	Flow[a] Without curing agent	With 5 phr EPIKURE 107FF
50	G	4.0	1	2
	None	—	7	4
	B	0.5	7	4
100	G	4.0	8	2
	G	2.0	9	2
	F	0.3	9	5
	None	—	10	5
	B	0.5	11	7

[a]Flow assessment: 1 = excellent; 4–5 = good; 8 = poor.

TABLE 4.6
Effect of Curing Agent on Powder Melt Viscosity

	EPIKOTE 1055 TiO$_2$ Modaflow	EPIKOTE 1055 TiO$_2$ EPIKURE 107FF	EPIKOTE 1055 TiO$_2$ Modaflow EPIKURE 107FF
Melt viscosity at 0.221 s^{-1} and 150°C, Pa.s	4000	15	58

Component concentrations: EPIKOTE 1055   100 pbw
                           TiO$_2$ (Kronos CL 220) 100 pbw
                           Modaflow   0.5 phr
                           EPIKURE 107FF   5.0 phr

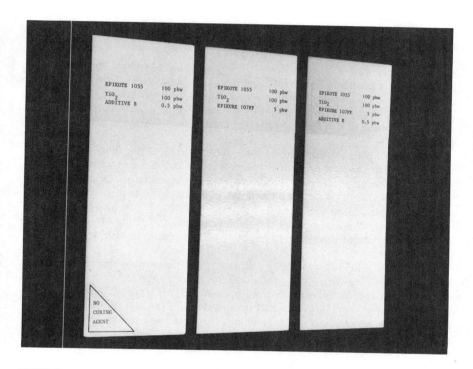

FIGURE 4.14   Influence of curing agent on flowout of powder coatings.

1) The Weissenberg Rheogoniometer can be used to
determine the yield values of such systems.
2) Yield values (a) increase with increasing pigment
concentration; (b) remain unchanged on addition of
acrylate anticratering additives; and (c) can be
strongly reduced by the addition of flow-promoting
additives.
3) Yield values are therefore a measure of the degree
of pigment dispersion and can be used to estimate this
property. The yield value, and hence the degree of
pigment dispersion, plays a large part in determining
the extent of flowout. Excellent flow can only be
achieved with systems having a yield value less than
3 Pa.

The addition of a curing agent to such systems can dramatically
lower the melt viscosity by functioning as a dispersion aid, and can therefore impart improved flow to the coating.

COMBINATIONS OF FLOW-PROMOTING AND
ANTICRATERING AGENTS

We have shown how improved flow in powder coatings can be achieved by a judicious choice of anticratering agent. We have also seen that even better flow can be obtained by the addition of a flow-promoting agent, but that the excellent flow is marred by the presence of cratering. Therefore, attempts were made to eliminate the cratering while maintaining the very good flow.

First of all the influence of the concentration of the flow-promoting additive G was examined. Levels between 0.5 and 8.0 phr were incorporated into powder coatings. Good flow was observed in all cases, but it was found that additive G is only partially effective against cratering at lower concentrations; levels of 6-8 phr were necessary to completely eliminate cratering. However, such concentrations cannot be used in practice because of the cost and the marked reduction in storage stability which results.

Further work was therefore aimed at finding suitable mixtures of additive G with known anticratering agents. Anticratering additives A to E were used in various concentrations in combination with additive G. In all cases, however, the flow obtained was equivalent to that obtained using the anticratering agent alone. Thus, it appears that even the relatively small reduction in surface tension produced by acrylate E, which has the lowest molecular weight, reduces the driving force for flowout to such an extent that pigment-wetting promoters cannot bring about an improvement in flow. However, these results were obtained with systems cured with nitrogen-containing curing agents such as EPIKURE 107FF, which function as potential pigment-wetting promoters. It could well be that in a system containing a catalyst or co-curing resin which is a less efficient pigment-wetting promoter, flow improvements could be obtained with combinations of anticratering and flow-promoting additives.

CONCLUSIONS

In general, the conclusions of our study are as follows:

1) Low molecular weight acrylate additives give good flow and are effective against cratering.
2) In uncured coatings, it is possible to determine the yield value which must be overcome before flow can take place.
3) Really good flow can only be obtained when this yield value is less than approximately 3 Pa.

4) A low yield value can be achieved by addition of a flow-promoting additive to improve the degree of pigment dispersion.
5) The curing agent can significantly influence the degree of pigment dispersion.
6) In the present study, combinations of flow-promoting and anticratering additives did not result in improved flow.

ACKNOWLEDGMENT

The author wishes to express his thanks to Mr. J. Raadesen for his contribution to this work.

REFERENCES

1. S. Gabriel, J. Oil Colour Chemists Assoc., 59:52, 1976.
2. P. G. de Lange: Proc. 6th Inter. Conf. in Org. Coatings Sci. and Technol., Vol. 4 - Advances in Organic Coatings Science and Technologies Series, Technomic, 1982.
3. J. F. Rhodes and S. E. Orchard: J. Appl. Sci. Res. Sect. A, II:451, 1962.
4. F. J. Hahn, J. Paint Tech., 43(562):58, 1971.
5. J. van Oosterom: Polymers, Paint and Colour J., 774, Sept. 6, 1978.
6. G. P. Bierwagen, Progr. Org. Coatings, 3:101, 1975.
7. W. A. Zisman and H. W. Fox: J. Colloid Sci. 5:514, 1950.
8. W. A. Zisman J. Paint Tech., 44(564):42, 1972.
9. W. J. McGill J. Oil Colour Chemists Assoc. 60:121, 1977.
10. E. J. Helwig J. Paint Tech. 41(529):139, 1969.
11. W. T. M. Johnson Official Digest 1489, November 1961.
12. W. K. Asbeck J. Coatings Tech. 49(635):59, 1977.
13. Amsterdam Monthly Technical Notes, 1: February 1977 (Available from the author on request).
14. USP 4 068 039 to W. R. Grace & Co., New York, 1978.

SOME ASPECTS OF INFLUENCING THE PROPERTIES
OF ORGANIC PIGMENTS BY USING
MODERN FINISHING METHODS

Klaus Hunger and Kurt Merkle

Hoechst AG
Frankfurt am Main, West Germany

INTRODUCTION

Organic colorants can be divided into two categories, namely dye-stuffs and pigments. Dyestuffs are by definition soluble in the application medium. They are either physically-absorbed or chemically-bound to the substrate to be colored. Examples of absorption are provided by direct dyes and of chemical bonds by reactive dyes. Such important coloristic properties of dyestuffs as shade, tinctorial strength, chemical resistance, and lightfastness are consequently determined mainly by their molecular structure. Only a few applicational properties, for instance solvent fastness, can be partially influenced by the degree of absorption in the vehicle. Sometimes a certain degree of influence on the lightfastness can also be observed as a function of substrate and application method. Polyester fibers, for example, thoroughly dyed by disperse dyes give much better lightfastness than the same fiber only partially dyed (ring-dyeing). But it can generally be stated that the properties of organic dyestuffs are mainly determined by the chemical structure of the dyestuff molecule and that the development of a dyestuff is essentially completed by the chemical synthesis.

A certain exception must be made with disperse dyes. Physically, they are positioned between dyestuffs and pigments as regards their application. They can therefore be considered as pigments with unsatisfactory solvent fastness. Disperse dyes are finally dissolved in the fiber as soon as the substrate is dyed.

Organic pigments on the other hand are insoluble in the substrate to be colored. They therefore exist only in crystalline form. Their applicational properties are thus essentially determined by physical aspects: their crystal structure; crystallinity; particle size distribution; and the condition of the crystal surface, and optionally the crystal modification. The properties of a pigment can be varied within a fairly broad range.

Apparently there is a relationship between the pigment properties and the chemical structure, since this factor in particular decisively determines the crystal structure and the crystallinity of a pigment. Also the basic shade of a pigment is, with a few exceptions, mainly defined by the chemical constitution. Although organic pigments are insoluble in the application media by definition, they show a more or less pronounced solubility in certain solvents or dissolving systems.

CHARACTERISTICS OF ORGANIC PIGMENTS

The facts discussed above render it possible to use special finishing methods to the manufacture of pigments. In order to illustrate this, some facts should be demonstrated taking as an example a quinacridone pigment— more precisely a linear trans-quinacridone. This is Pigment Violet 19. Figure 5.1 shows the chemical formula of the unsubstituted linear trans-quinacridone. At first sight, it can be seen that quinacridone is a rather small and not very complicated molecule although its production necessitates some extensive synthesis steps. Based on its chemical constitution, nobody would expect this quinacridone molecule to exhibit an intensive color. Actually the dissolved quinacridone molecule shows only a weak fluorescent-orange color. But in the crystalline state, quinacridones are intensively colored. This substance exists in various crystal modifications which differ in their properties. The α- and γ-modifications are red,

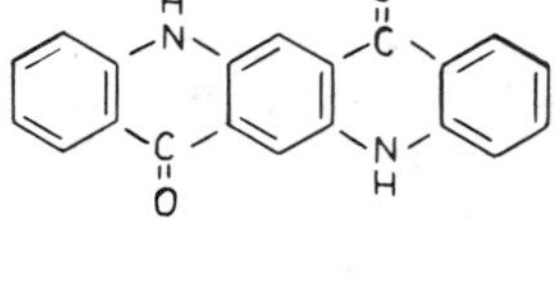

Pigment Violet 19
(46500)

FIGURE 5.1

the β-modification is an intensive red violet. The β- and γ-modifications of quinacridone are commercially used, contrary to the α-phase which has inadequate lightfastness. Figure 5.2 demonstrates the X-ray diffraction spectra of the different modifications. A crystal lattice quite similar to that of graphite, which causes the intensive color of the crystalline state of the crystalline state of quinacridones is discussed [1]. In any case, the interactions between the pigment molecules in the crystal lattice decisively influence the pigment color. This example has been shown to illustrate how considerable the influence of the crystal structure can be on such important pigment properties as tinctorial strength, shade, and lightfastness [2].

The existence of different crystal modifications for one and the same chemical individual is called polymorphy. Polymorphy is not limited to such well known classes of organic pigments as quinacridones and phthalocyanines. Many other organic pigments of different chemical classes are polymorphic, although in these classes scarcely more than one crystal modification is being used commercially. Figures 5.3 and 5.4 demonstrate polymorphy of two different azo pigments: one of the benzimidzolone type, the other one of the disazo type.

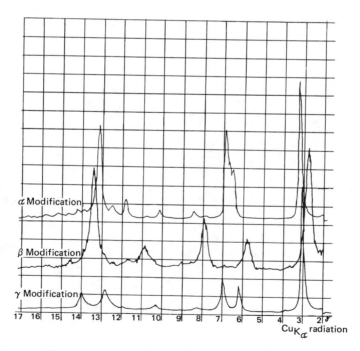

FIGURE 5.2   X-ray diffraction spectra of the different modifications of quinacridone pigments.

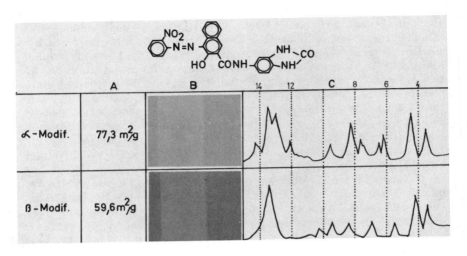

FIGURE 5.3   Monoazo pigment
               A: Specific surface area (BET)
               B: Graphic printing 10%, the middle part exposed to light
               C: X-ray diffraction spectra

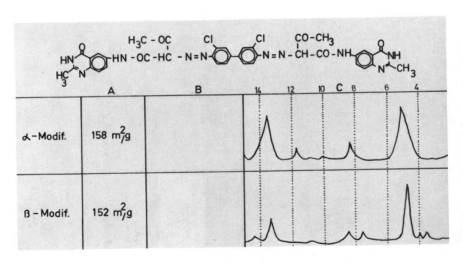

FIGURE 5.4   Disazo pigment
               A: Specific surface area (BET)
               B: Graphic printing 10%, the middle part exposed to light
               C: X-ray diffraction spectra

It is known from many patent applications of producers of organic pigments that the desired crystal modification of a polymorphic pigment can be obtained by employing special finishing conditions. These include the use of certain solvents and surfactants, the time, and the temperature of the thermal treatment. Methods and conditions have to be established for each individual pigment and cannot be transferred from one to another. Generally speaking however, obtaining the specific crystal modification required in a polymorphic pigment no longer presents a very big problem with the modern finishing methods available. On the other hand, it is much more difficult to impose the crystal modification of a certain pigment upon a pigment with a different chemical constitution.

The fact that not only the chemical constitution but also the corresponding crystal modification influences the coloristic and fastness properties of an organic pigment offers some interesting aspects concerning the required improvement of the properties of simple organic pigments. Based on this consideration it should be possible to improve a cheap pigment of relatively simple constitution and low-fastness properties by attempting to have it crystallized in a modification which belongs to a pigment with more valuable properties.

A method occasionally used in past years consists of the so-called mixed coupling process of azopigments. The addition of only a small amount of a second diazo or coupling component to a coupling reaction enables one in some cases to achieve a desired and distinct improvement in the properties of the parent pigment. This can be demonstrated with the following example: Pigment Yellow 12 and 13—two diarylide pigments, each with 3, 3'—dichlorobenzidine as the diazo component, differ only by virtue of substitution pattern of the aromatic nucleus of the acetoacetic arylide (Figure 5.5).

Pigment Yellow 12 is based on the cheaper acetoacetic anilide, which can be synthesized from diketene and aniline, has a greenish yellow shade, lower lightfastness, and tinctorial strength compared with the more expensive Pigment Yellow 13, which contains acetoacetic metaxylidide, made from diketene and m-xylidine, as coupling component.

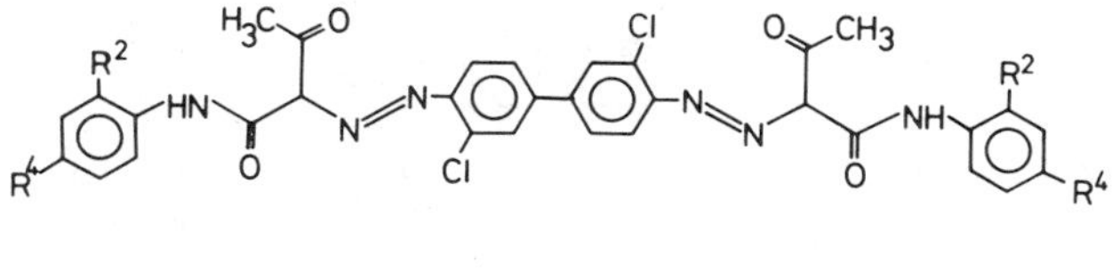

R^2, R^4 = H : Pigment Yellow 12

R^2, R^4 = CH_3 : Pigment Yellow 13

FIGURE 5.5

By a mixed coupling reaction of bisdiazotized dichlorobenzidine with a mixture of acetoacetic anilide and about 10% of acetoacetic p-anisidide (the latter is Pigment Yellow 170) (Figure 5.6) one obtains under specific conditions a pigment which, as expected, consists of more than 90% of Pigment Yellow 12 (I) and a pigment mixture which contains less than 10% of the acetoacetic p-anisidide rest. Because of the two functional sites of bisdiazotized dichlorobenzidine the acetoacetic p-anisidide is built into the symmetric (III) and the asymmetric pigment molecule (II).

But the important aspect of this synthesis is that the mixed pigment has the better coloristic and fastness properties of the acetoacetic m-xylidide pigment Pigment Yellow 13, although it is crystallized in the crystal modification of the anilide pigment Pigment Yellow 12 (Figures 5.7-5.9). This is shown with X-ray powder spectroscopy.

Whereas the formation of a desired crystal modification obtained by alternative finishing methods is in some cases possible, it is much more problematic to obtain a specific particle size distribution. It is even more difficult to reach this goal for pigments with good solvent fastness.

The particle size distribution of a pigment is a very important factor, since it is not only responsible for the coloristic but also for the applicational properties of the pigment, since besides shade and brilliancy dispersibility, rheological behavior in the vehicle, hiding power, dispersion stability, and fastness properties are concerned [3].

Some inorganic heavy metal containing pigments which are included in several pigmented vehicles have recently had to be replaced by organic pigments. Since inorganic pigments with high opacity were chiefly concerned, the pigment manufacturers called for organic pigments with high hiding power. The hiding power of a pigment is influenced by its refractive

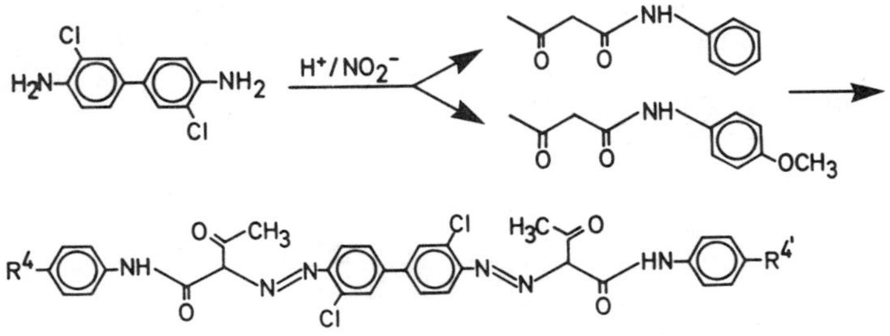

I : R⁴, R⁴' = H
II : R⁴ = OCH₃, R⁴' = H
III : R⁴, R⁴' = OCH₃

FIGURE 5.6   I: Pigment Yellow 12
              III: Pigment Yellow 170

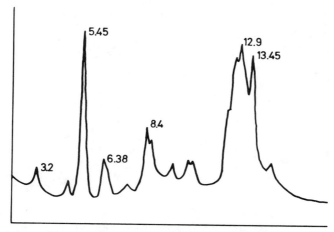

FIGURE 5.7   X-ray diffraction spectra of Pigment Yellow 170

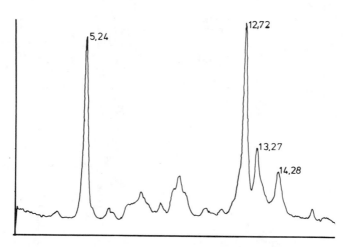

FIGURE 5.8   X-ray diffraction spectrum of Pigment Yellow 12.

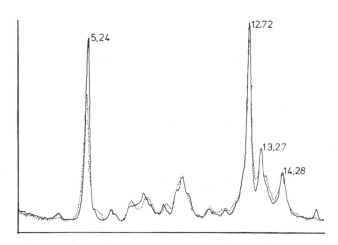

FIGURE 5.9   X-ray diffraction spectra-comparison of P. Y. 12 (through
line) with the mixed coupling product P. Y. 12/P. Y. 170
(dotted line).

index or more precisely by the difference between the refractive indices
of a pigment and the binder, by the scattering coefficient and the pigment
concentration.   The refractive index is a material constant and cannot be
influenced or changed by finishing methods.   The refractive index of
organic pigments is, depending on the wavelength, about 1. 2 to 2. 5.   For
several important yellow, orange, and red pigments, the refractive index
is about 1. 5-1. 6, so that in paints whose binders have a refractive index
of about 1. 5, a not very high opacity can be expected.   An increase in the
hiding power of organic pigments can therefore only be achieved by increasing absorption [4] and scattering.   Pigments with a high absorption coefficient such as carbon black or dioxazine violet permit the production of
layers with high hiding power, although they are highly transparent.

The reason for this effect is the strong light absorption of the deep
shade.   The light absorption of a red pigment with medium absorption contributes only to a certain extent to hiding power.   The influence of absorption on opacity decreases further with increasing shift of the shade towards
yellow.   An improvement in opacity of a yellow or red shade can therefore
only be achieved by increasing the light scattering.   This can be done by
changing the particle size distribution.

For an understanding of the increase of light scattering by particle-
size distribution, it is necessary to know about the theory of Mie [5], which
shows that maximum light scattering of small spherical particles can be
calculated as a function of their diameter.   It is now possible to calculate
the optimum light scattering even for nonspherical pigment particles by the
extended theory of Mie with the aid of modern computer programs.

With such a calculation, one can determine at which medium particle size the optimum of light scattering and subsequently the optimum opacity can be obtained. With the rule of thumb out of the Mie theory, the optimum light-scattering effect can be expected at a particle size about equal to half of the wavelength of the light absorbed at its absorption maximum. Figure 5.10 shows the different particle size distribution of Pigment Orange 34: the standard type (I) and the type finished up to its maximum opacity (II). Molybdate Red is shown for comparison (III). For reaching the optimum opacity, it is necessary to control the process with regard to the particle-size distribution in order to obtain an optimum number of particles with a mean diameter as close as possible to the precalculated value.

Figure 5.11 shows the calculated particle size for the range of optimum scattering with 0.23 μm, the mean average of the technical grade Pigment Orange 34 in its most opaque form is 0.21 μm. The common type of Pigment Orange 34 has a mean average size of 0.07 μm. In Figures 5.12 and 5.13, the difference of hiding power between the standard commercial Pigment Orange 34 and the type which has been finished to its maximum opacity is shown. Figure 5.12 shows the difference for equal concentration of pigment weight and Figure 5.13 for equal pigment-surface concentration.

Because of the different specific surface areas of the compared types of this pigment (Figure 5.14) different pigment volume concentrations are also obtained. It can generally be stated that an increase in light scattering for equal pigment-volume concentration results in a distinct increase in the hiding power of the pigmented system.

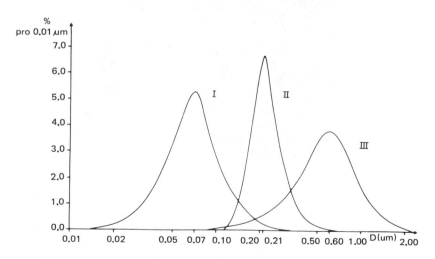

FIGURE 5.10   Particle size distribution: I, P. Or. 34, commercial grade; II, P. Or. 34, opaque type; III, Molybdate red.

Particle Size [μm]

Pigment	Mean Value	Optimum of Scattering [μm]
Pigment Orange 34	0,07	
Pigment Orange 34 Opaque	0,21	0,23
Organic Pigments in general	0,05 – 0,5	
Lead Chromates	0,6	

FIGURE 5.11

FIGURE 5.12 Pigment Orange 34 over black and white cardboard.
Pigment concentration by weight 0.1%.

FIGURE 5.13 Pigment Orange 34 over black and white cardboard.
Pigment surface concentration 5.0.

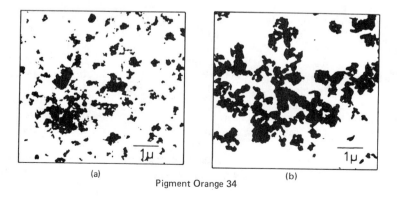

(a)

Pigment Orange 34

(b)

FIGURE 5.14 (a) Commercial grade; (b) opaque type. Specific surface
area: 48.4 m^2/g and 13.6 m^2/g, respectively.

FINISHING METHODS

Let us now shed some light on the question of how a desired particle-size distribution in an organic pigment can be obtained with modern finishing methods. To start with one has to differentiate between three situations. A pigment obtained after its synthesis is:

(a) in more or less large crystals, which have to be finely divided by mechanical means;
(b) in small crystals which possess a small but limited solubility in certain solvents; or
(c) in very small particles which have only very low solubility in solvents.

(a) involves known organic pigments of the phthalocyanine, quinacridone, perylene, and naphthalene tetracarboxylic acid type. These pigments have to be ground mechanically, for instance with sodium chloride or sulfate or other substances as grinding auxiliaries, in order to produce the desired particle size distribution. Grinding time, use of the requisite milling aditives, and the type of grinding technology are responsible for specific decrease in the mean particle size. This sounds simple. But as a rule, it needs extensive experiments in order to get a certain desired optimum. All parameters obtained in a series of experiments are only valid for this distinct pigment or modification.

(b) involves azopigments of the arylide, pyrazolone, and ®Naphtol AS type. Azo pigments are obtained after the coupling reaction in a very small mean-particle size. These pigment crystallites are smaller than 0.1 μm and thus are within the scope of validity of the Thompson-Gibbs law regarding solubility in the colloidal range. The solubility of such small crystallites increases approximately with the second power for decreasing particle size. It means for instance that a bisection of a certain particle size increases the solubility four times. If a particle is reduced to one third of its original size, its solubility is increased by the factor 9.

This fact renders it possible to increase the mean particle size of a pigment to the desired degree by a solvent treatment at elevated temperatures. In simple cases, water can be used instead of organic solvents. This thermal treatment is generally understood as a solvation process. A saturated solution for small particles means, according to Thompson's law, a strong supersaturation for larger crystallites. Consequently, larger particles tend to grow at the expense of smaller ones. This is valid for the typical particle size distribution of organic pigments and cannot be applied for one distinct particle size, which might at least be theoretically possible.

The finishing process can be controlled by the choice of the solvent, the pigment concentration, and the temperature. The rate of diffusion can be influenced by mechanical stirring. These factors have to be taken into consideration in order to approximately reach the desired particle size distribution. The finishing process together with the decrease in the specific surface area causes a decrease in the agglomeration tendency which consequently means a narrower particle-size distribution. The more contracted the particle-size distribution curve is the cleaner the shade of the pigment.

The process of solvent finishing of a pigment sometimes takes place unintentionally during manufacture, shearing forces of the pigment dispersion may cause adiabatic heating. Smaller pigment particles will consequently be dissolved and grow on the surface of the less soluble larger particles. This so-called recrystallization process is chiefly known for pigments with low solvent fastness. Prerequisites are: the vehicle which should act as protective colloid does not have enough of a stabilizing effect; the solvent used has high solvation power; and elevated temperatures are generally used. These facts can lead to an uncontrollable finishing process with the consequence of a broad particle-size distribution creating a duller shade, loss of tinctorial strength, and loss of transparency or opacity. However, pigment manufacturers have in numerous cases succeeded in obtaining high recrystallization stability for pigments with low solvent fastness by means of suitable surface treatment and the addition of special stabilizers [6].

Finally, with regard to (c), the most difficult finishing process concerns pigments which after synthesis are obtained in a very small mean-particle size and which have a relatively high solvent fastness. Pigments which belong to this type are for instance benzimidazolone and isoindolinone pigments. This type of pigment could be particularly valuable in its opaque form for high-quality industrial lacquering, for instance for automotive topcoat finishing and corresponding repair systems. Requirements for this application include very good light and weather fastness, and an excellent resistance to bleeding.

A contradiction exists between the properties of these pigments that are mainly desired: bleeding fastness, which means inertness to the solvation capability of solvents on the one hand; and the requirement for pigments with larger crystallites for more light scattering, which is necessary for high hiding power on the other. The contradiction is considerable, since the desired size of pigment crystallites for achieving an optimum scattering effect follows only to a very small extent the Thompson-Gibbs relation between solubility and particle size of crystals. Despite this fact, it was possible for pigments of the benzimidazolone type (Figure 5.15) and also some specially selected Naphtol AS pigments which belong to type (c) to be treated successfully in order to satisfy these requirements and obtain a reasonable compromise.

On radiation of pigments with light, photons have only a very small depth of penetration of about $0.03-0.07$ μm into the crystallite as has been calculated by Eulitz [7]. Consequently, the increase in size of crystallites leads to a considerable improvement in light- and weatherfastness. The

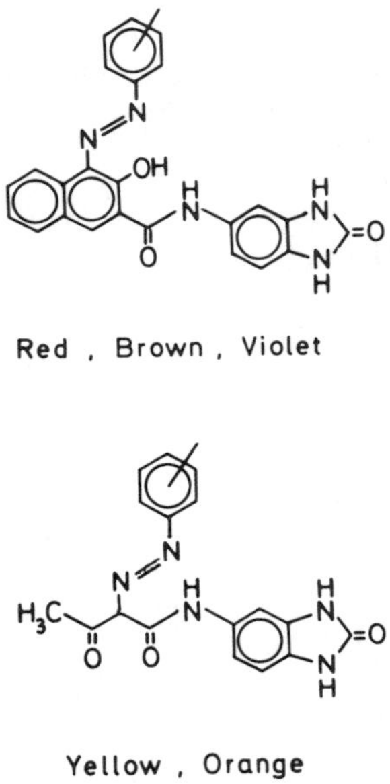

Red , Brown , Violet

Yellow , Orange

FIGURE 5.15   Benzimidazolone pigments.

destruction of a pigment by light proceeds layer by layer; this means that a large crystallite has a longer life than a small one. However, the light and weather fastness cannot be exclusively explained by the increase in size of the pigment crystallites. Another influence which explains the improvement in light and weather fastness by a finishing procedure is the better crystallinity, which means better-formed pigment crystals after the treatment. An example for this is demonstrated on Figure 5.16 [8].

Figure 5.16 shows X-ray diffraction curves of the untreated and the treated Pigment Orange 34. One can clearly see the sharper peaks of the after-treated pigment which indicates better crystallinity. The fact that it is possible for pigments with very low solubility to be finished in the presence of organic solvents and to reach the desired goal as far as opacity is concerned, without adversely affecting their excellent solvent fastness, requires an explanation.

The compromise between the two conflicting properties can only be obtained since the relationship between solubility and temperature of these benzimidazolone pigments yields a curve which is quite different to that of other pigments with similarly good light and weather fastness (Figure 5.17).

The benzimidazolone pigments possess relatively low solubility at temperatures up to 160°-180°C. This behavior changes abruptly in this

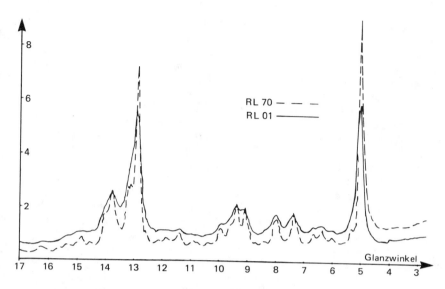

FIGURE 5.16 X-ray diffraction spectra of P. Or. 34
RL 01: Commercial grade
RL 70: Opaque type

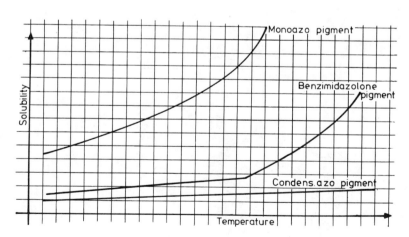

FIGURE 5.17 Solubility versus temperature for different types of pigments.

range when the temperature continues to rise. A noticeable amount of pigment is now being dissolved. In addition to the curve for benzimidazolone
pigments, the curves for monoazo pigments and for condensation azopigments are illustrated. Whereas monoazo pigments show a typical trend:
steadily increasing solubility with increasing temperature, the condensation
pigments remain almost insoluble throughout the whole temperature range.

The reason for the noticeable behavior of the benzimidazolone pigments lies in the high polarity of these pigment molecules caused by the
accumulation of several carbonamide bonds and, particularly, by the
benzimidazolone group. As is well known, carbonamide bonds are responsible for the formation of hydrogen bonds. Their break eventually
proceeds at the temperature of the sudden change of solubility. Polar solvents are preferred for this pigment treatment.

Recent results of a three-dimensional X-ray analysis of a benzimidazolone pigment support this assumption. All crystal structures of different kinds of azopigments known to us up to now show intramolecular
hydrogen bonds exclusively. This means all hydrogen bonds possible are
located in the single molecule (Figure 5.18) [9]. The complete determination of a red benzimidazolone pigment has proven our assumption. This
pigment is characterized by additional intermolecular hydrogen bonds.
This means bonds between different pigment molecules—a result completely
different from all azopigments known until now (Figure 5.19) [10].

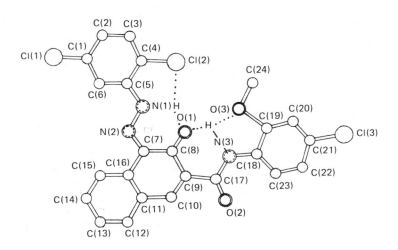

FIGURE 5.18 Chloro derivative of Pigment Red 9.
.... Intermolecular hydrogen bonds.

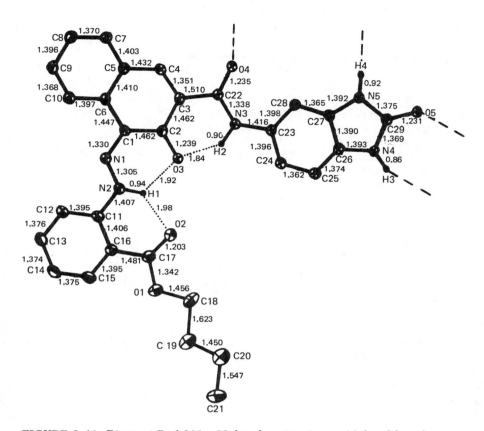

FIGURE 5.19 Pigment Red 208. Molecular structure with bond lengths.
Hydrogen bonds: .... intramolecular
                    ------- intermolecular

CONCLUSIONS

This paper was intended to present a view on some of the problems and possibilities of modern pigment-finish. In numerous cases, the applicational properties of organic pigments can thus be changed in a desired direction with the aid of these or other finishing methods. But despite the considerable progress made in this field, a great deal of practical experience is still necessary. In many cases, a more or less effective compromise has to be found by means of extended series of experiments. These facts should explain why producers of organic pigments keep their knowledge and experience on pigment finishes confidential as valuable know-how.

We believe that the colorants presently known in the field of organic pigments sufficiently satisfy most of the realistic requirements of pigment manufacturers. Therefore, the pigment finish, which involves the specific change in the properties of colorants already known, will gain much more importance in the future compared with the synthesis of new substances.

The economic situation and the relevant legislative requirements which have been arisen in recent years in almost all Western industrial countries will compel such development, since new products are going to need higher expenditure on market introduction and in particular since very extensive toxicological studies have to be performed for each new compound on the market.

REFERENCES

1.   Linke, G. Farbe und Lack 86:966, 1980.
2.   Herbst, W., and Merkle, K. Defazet 8:365, 1970.
3.   Herbst, W., and Merkle, K. Defazet 29:433, 1975.
     Herbst, W., Progr. Org. Coatings 1:267, 1972/73.
4.   Biffar, H. J. Farbe und Lack 8:1025, 1976.
5.   Mie, G. Annalen der Physik 25:377, 1908.
6.   Memmel, F. and Merkle, K. Plastverarb. 6:336, 1966.
     Herbst, W. and Hunger, K. Progr. Org. Coatings 6:105, 1978.
     German Pat. 2012152 and 2012153, Hoechst, A. G., 1970.
7.   Eulitz, G. private communication.
8.   Merkle, K., Herbst, W., and Eulitz, G. Farbe und Lack 9:801, 1976.
9.   Whitaker, A. J. Soc. Dyers Col. 431, 1978.
10. Paulus, E. F. and Hunger, K. Farbe und Lack 86:116, 1980.

PARTICLE SIZE ANALYSIS IN COATING SYSTEMS

P. J. Lloyd

University of Technology, Loughborough
Leicestershire, England

ABSTRACT

This paper reviews the methods of particle size analysis available for characterizing coating systems and explains why differences may be experienced in the particle-size distributions measured by different instruments or methods and why when using even one instrument, differences still occur because of dispersion difficulties, bad sampling, or intrinsically, because of the finite number of particles measured.

INTRODUCTION

No method of analysis is quite so confusing as that for particle size. A recent survey for the Particle Size Group of the Royal Chemical Society identified over 400 instruments or methods. Many of these were very similar but even today there are between 50 to 100 methods available.

A classification which enables some order to be made out of this complex situation is presented. The methods are classified by the method of making the measurement or the physical process when that is the most significant. Most methods are available in a number of commercially available instruments although where one manufacturer is particularly significant, that name is given. Also demonstrated are many different

particle properties; this is perhaps the first clue to one of the many con-
fusions in particle size analysis. Generally one expects a successful
method of analysis to yield a unique result and a background in chemical
analysis would lead one to expect that this result could be obtained by using
different routes or techniques. This is not so in particle size analysis as
the result obtained by one method may not agree with that obtained by another. Indeed the results of replicate analyses using the same instrument
may show a much larger spread than one would normally expect. The
reasons for these technical difficulties are many but can be described
under three principal headings: (1) physical properties, (2) chemical
properties, and (3) statistical properties. Each of these will be described
in full so that where differences in the results of analysis occur they can
be understood. These difficulties are the same whatever the particle size
range considered but in the range of 10 μm to 0. 01 μm the problem of
dispersion may be particularly severe.

PHYSICAL PROPERTIES

As indicated in Table 6. 1 different properties of particles are used
to identify particle size. Only if the particles are spherical can a unique
particle diameter be measured which will enable other properties to be
calculated since the formulas relating the diameter (d) to all the particle
properties are known, e.g. ,

Volume $\quad \dfrac{\pi d^3}{6}$ $\hfill (1)$

Projected area $\quad \dfrac{\pi d^2}{6}$ $\hfill (2)$

Surface/volume ratio $\quad$ 6/d $\hfill (3)$

Settling velocity $U_t = \dfrac{\rho - \rho_0}{18\eta}\, gd^2$ $\hfill (4)$

ρ and ρ_0 are the densities of particle and medium respectively; η is the
viscosity of the medium. But even for other simple well-defined shapes,
the particle size cannot be described by a single parameter, for example,
rectanguloid needs three parameters. Even with this information it is not
possible to calculate all the particle properties. Thus, there exists a
problem even with these simple shapes which becomes much greater with
more irregular particles.

In particle size analysis, a pragmatic solution is taken. For a
sphere, most of the physical properties can be calculated. So if any

TABLE 6.1
A Classification of Particle-Sizing Methods

Classification	Particle-sizing method	Particle property measured
Field scanning	Optical microscope Transmission electron microscope Scanning electron microscope Diffractometers Turbidity measurement/ nephelometers	Particle cross-section area
Classification	Sieving Air classification Elutriation	Cross-section dimensions Hydrodynamic Hydrodynamic
Stream scanning	Electrical sensing zone (Coulter counter) Light scattering High blocking (Hiac counter)	Volume Cross-section area Cross-section area
Sedimentation	Andreason pipette Sedimentation balance Photosedimentometers X-ray and γ-ray sedimentometers Centrifugal sedimentometers	Hydrodynamic
Surface area	Permeametry Gas adsorption Dye adsorption	Surface/volume diameter
Brownian motion	Photon correlation microscopy	Diffusion coefficient

physical property of a particle can be measured, the diameter of a sphere with the same property can be calculated. This diameter is termed the "particle size" or the equivalent sphere diameter. This has the effect of making the particle size dependent on the method used to measure it and so it should not be expected that the particle size measured by one method will be similar to that measured by another. Indeed, the sizes will only coincide when the particles are spheres and the farther the shape of the particle diverges from that of a sphere the wider will be the discrepancy. Figure 6.1 shows how the equivalent sphere changes as different properties are considered. It is seen that for the rectanguloid, the differences illustrated are quite large.

Although this concept does overcome the problem of defining particle size and enable its measurement, it does presuppose that the conversion of the particle property into a single equivalent sphere diameter is always readily accomplished. This may be too simple an idea as the following two examples will demonstrate.

Example 1: The use of equation (4) to convert the settling velocity of a sphere to its diameter assumes the universality of the equation. In fact, even for spheres the equation is only true over a limited particle size range. The upper size limit is set by the condition that the Reynolds Number $Re_p = \rho\,d\,U/\eta < 0.2$ which is only true for particles of density 2500 kg/m^3 and diameter $< 50\,\mu$m. The lower limit is less well-defined. But for particles of about 1 μm, the sedimentation velocity will be modified slightly by Brownian diffusion and the magnitude of the effect will be larger as smaller particle sizes are considered. A further source of error occurs because equation (4) is only true for particles is a very dilute suspension. At higher concentrations than 0.05 vol/vol % the settling velocity will be affected and at high concentrations 1-10 vol/vol % the velocity will be independent of particle size by a function only of concentration.

Example 2: Concentration can also lead to errors in the size analysis obtained in a Coulter Counter and similar electrical zone counters. Recent papers by Lloyd [1] and Harfield and Knight [2] show conclusively that for dilute suspensions the response of the instrument is proportional to the volume of the particle. Thus, if the instrument is calibrated using spheres, it will measure the equivalent sphere diameter with the same volume of the particle. However, if more than one particle is in the sensing zone at the same time the response is very complex and can lead to the loss of counts and recording the pseudo-particles of equivalent volume between that of a single particle and that of the combination. This intrinsic error can be avoided by using a dilute suspension and testing that the size distribution obtained is independent of concentration.

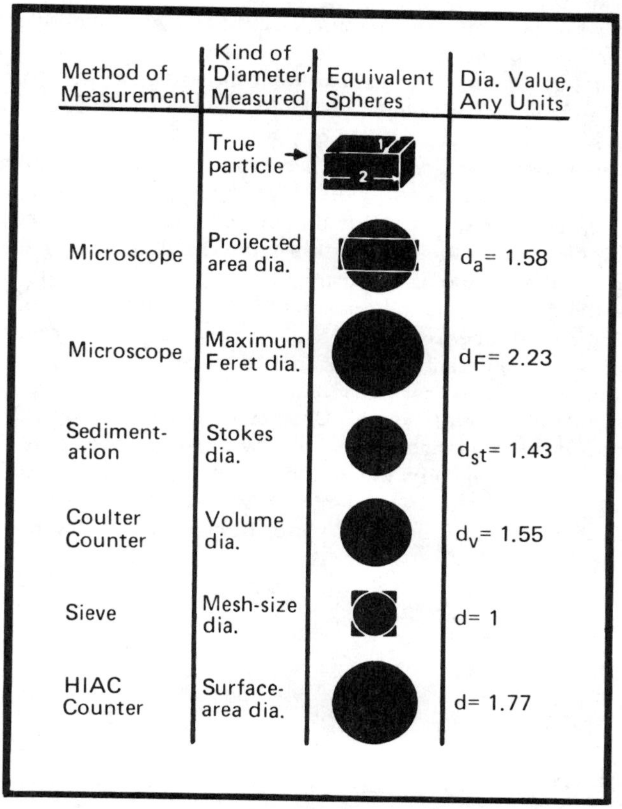

Method of Measurement	Kind of 'Diameter' Measured	Equivalent Spheres	Dia. Value, Any Units
	True particle →		
Microscope	Projected area dia.		$d_a = 1.58$
Microscope	Maximum Feret dia.		$d_F = 2.23$
Sediment- ation	Stokes dia.		$d_{st} = 1.43$
Coulter Counter	Volume dia.		$d_v = 1.55$
Sieve	Mesh-size dia.		$d = 1$
HIAC Counter	Surface- area dia.		$d = 1.77$

FIGURE 6.1

CHEMICAL PROPERTIES

Most particle size analysis instruments require a dilute suspension of the particles. A dry powder has to be dispersed in a liquid or a concentrated solution has to be diluted. The usual practice is to add dispersing agent and use an ultrasonic bath to complete the dispersion. Although this is common practice it should not be followed without some care. It is possible that incomplete dispersion may occur and the size analysis obtained be solely a function of the dispersion agent used, its concentration, and the time interval between the dispersion and the analysis. The magnitude of this effect was demonstrated by Lloyd et al. [3]. The powder was a feldspar and transmittance of a parallel light beam was used as a measure of the dispersion. As a powder is dispersed more surface area is

released which reduces the amount of light transmitted—thus, the lower
the transmission, the higher the dispersion. Figure 6.2 shows the results
obtained using Nonidet P42 and two concentrations of the recommended
dispersing agent for a feldspar powder. These results are interesting
since they show the unsuitability of the Nonidet P42 in that the transmittance is increasing with time which can only be caused by the agglomeration
of the particles. The results with the sodium pyrophosphate show that the
high concentration causes better dispersion and in each case the dispersion
is not complete until 10–100 minutes after the start. It is thus demonstrated that both the choice of dispersing agent is important and the concentration at which it is used are important. Fortunately, lists of recommended
dispersing agents are available [4]. But even if the correct agent is
selected, it is important to check that the concentration at which it is used
is the correct one and that sufficient time is left between dispersion and
analysis. Results obtained on a Coulter Counter (Table 6.2) demonstrate
how counts in each channel show a marked change over a series of measurements. These changes mean that effectively the size distribution is a
function of time; this does not lend confidence in the results. Clearly, in
any analytical situation it is important to realize that the particle size

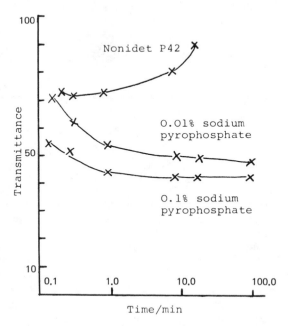

FIGURE 6.2   Variation of transmittance as function of time with different
dispersing agents.

TABLE 6.2

Distributions Obtained in a Coulter Counter Over 20 Min

Channel	Count									
0	8	4	4	6	3	0	2	6	2	5
1	7	7	11	7	3	8	8	7	6	11
2	95	86	64	67	61	51	48	55	48	40
3	717	547	476	458	372	342	324	300	289	265
4	1470	1286	1134	1005	941	840	773	718	612	619
5	749	681	602	581	537	442	423	364	333	378
6	858	707	682	659	580	509	527	470	444	449
7	903	820	822	710	712	642	643	580	557	565
8	1220	1240	1115	1045	1120	1015	1012	953	863	876
9	2048	1869	1900	1775	1823	1700	1634	1602	1482	1523
10	3726	3531	3552	3363	3467	3331	3187	3272	3094	3034
11	7869	7390	7398	7174	7266	6922	6764	6912	6653	6636
12	17505	17191	17119	16734	16976	16792	16330	16363	15973	16085
13	52101	51686	51763	50683	51271	50577	50269	50646	49601	50584
14	98353	93683	97167	93272	95964	92165	91865	94488	91415	93570

distribution should not be dependent on concentration, time, or the amount of dispersing agent used.

Even greater care must be taken when the sample to be analyzed is a concentrated suspension. Usually additives are present to ensure that the suspension of particles or emulsion remains stable. Dilution to concentration suitable for analysis may cause flocculation and completely alter the size distribution. There is no easy answer if it is required to obtain a size distribution. The only satisfactory method is to smear the concentrated suspension on a microscope slide and then to manually count the particles. It is only in this way can the possibility of destroying the structure with dilution be overcome.

STATISTICAL PROPERTIES

Even in a correctly-prepared dispersion it may be found that the size distributions obtained in a series of analyses are not identical. There are two possible reasons for this. First, there is an intrinsic error due to the measurement being made on a finite number of particles, and second there may be a sampling error due to inaccurate subdivision of the sample.

The British Standard on Particle Size Analysis [5] and more recently Lloyd et al. [3] have shown that each point N(x) in a particle size distribution has an instrisic error σ_R given by

$$\sigma_R = \sqrt{\frac{N(x)\ (1 - N(x)\ )}{n}} \tag{5}$$

where N(x) is the cumulative number fraction of particles less than size x, and n is the total number of particles counted. This equation clearly demonstrates the dependence of the intrinsic error on the number of particles counted. The expected range of values of N(x) for 95% confidence is $N(x) \pm 2\sigma_R$ and Table 6.3 shows the magnitude for different numbers of particles counted.

These errors are significant and show the necessity for counting reasonably large numbers of particles. The minimum number of particles recommended in BS 3406 [5] is 625 which yields an intrinsic error of ± 0.04 on the mid point of the distribution (N(x) = 0.5).

Even so the actual error may be much larger if the original sample is not obtained properly. It is too easy to take a small quantity of the powder with a spatula and add it to the liquid plus dispersing agent. This is bad practice and will lead to error. Every particle in the original quantity of powder must have an equal opportunity to appear in the sample for analysis. This means using a commercial sample divider or more simply, cone and quartering. The powder is made into a heap and is divided into four quarters. Diametrically opposing quarters are discarded

TABLE 6.3
Errors Due to Number of Particles Counted

N (x)	n=100	n=1000	n=10,000	n=100,000
0	0.00	0.00	0.00	0.00
0.1	0.06	0.019	0.006	0.0019
0.2	0.08	0.025	0.008	0.0025
0.3	0.092	0.029	0.009	0.0029
0.4	0.098	0.031	0.009	0.0029
0.4	0.098	0.031	0.0098	0.0031
0.5	0.100	0.032	0.010	0.0032
0.6	0.098	0.031	0.0098	0.0031
0.7	0.092	0.020	0.009	0.0029
0.8	0.06	0.019	0.006	0.0019
1.0	0.00	0.00	0.00	0.00

and a new heap is made with the remainder. This is repeated until the
sample size is small enough. Even with this amount of care the between
sample standard deviation may be much larger than normally expected in
analysis, in practice approximately three to four times the value of σ_R in
equation (5).

PARTICLE SIZE ANALYSIS FOR COATING SYSTEMS

Methods available for particle size analysis fall naturally into two
classes: those above one micron and those below one micron. The above-
one micron methods are many in number and some can be used to as low a
size as 0.1 μm while there are only a few available to measure particles
below one micron. These methods are reviewed in the classification introduced earlier and comments are made as to the applicability for measurement of particle size in coating systems.

FIELD SCANNING—MICROSCOPY

One common feature to all microscope methods is that it is very
time-consuming to count a large number of particles, so the distribution
may be subject to quite large errors on each distribution point. Great
care has to be taken to ensure that the particles counted are representative of the total quantity.

Optical Microscope	Range 100 - 1 μm
Scanning Electron Microscope	Range 10 - 0.1 μm
Transmission Electron Microscope	Range 1 - 0.01 μm

Automatic microscopy can increase the number of particles counted and currently available software enables any equivalent size to be measured. However, great care is necessary to prepare a slide such that the particles are all individual, i.e., are not touching, and of sufficient contrast for the instrument to be able to detect boundaries easily.

FAR FIELD DIFFRACTION

In recent years, three commercial instruments have been marketed and are proving very valuable. The far-field diffraction pattern is obtained by shining a laserbeam onto a suspension of particles and is recorded by measuring the intensity of the light as a function of angle and scatter using photoconductive cells. Using a small digital computer, which is part of the instrument, these data are converted into a size distribution. The procedure adopted differs between the three instruments but can be summarized as follows. The intensity at the largest angle can only be due to the smallest particles in the distribution. At smaller angles the intensity due to particles already considered is subtracted from the recorded intensity and the difference is assumed to be due to the size which will diffract the light to that angle. Errors can occur if there are smaller particles present which diffract light to larger angles than that detected and also if very large particles are present which diffract the light to angles too small to be detected. However, these possible errors do not detract from what has become a very valuable method. The size range is 1000 - 1 μm.

BROWNIAN MOTION

One of the most exciting developments in sub-micron sizing is the introduction of photon correlation microscopy. Particles suspended in dilute suspension move randomly under the influence of Brownian motion. The rate of movement is inversely proportional to the particle size. If the particles are illuminated by a low-power laser, the light scattered by the moving particles causes intensity fluctuations which can be used to give a measure of particle size. The particles are assumed to be spheres and currently instruments will give a mean particle size and an ill-defined dispersity index. This method is widely used in quality control situations where the size range has previously been very difficult to cover. It is hoped that, in the near future, more complete particle-size distributions will be obtained by this method. Statistics are good because observations are made on large numbers of particles. The size range is 3 - 0.02 μm.

STREAM SCANNING

Electrical Zone Sensing (e. g. , Coulter Counter)

Since 1945 the Coulter Counter and other instruments based on this principle have been used to size particles. The principle essentially is to measure the change in resistance between two electrodes as the particle passes through a small orifice which is the only conducting path between the electrodes. The change in resistance is recorded as a voltage pulse caused by the instrument maintaining a constant current between the electrodes. The change in resistance is recorded as a voltage pulse caused by the instrument maintaining a constant current between the electrodes. These instruments are very popular despite the requirement that the particles have to be dispersed in a conducting liquid. The range of application is large and measurement down to 1 μm can be considered routine and with great care it is possible to measure below this limit. Large numbers of particles can be counted in a very short time and so the statistical problems described earlier are easily overcome. The size range is 1000 - 0.5 μm.

Light Scattering (e. g. , Royco Counter)

Over a limited range, the light scattered from an individual particle at a preselected angle can be used to size the particle. The assumption of a spherical particle is used and a very dilute suspension is required. The principle is usually used in aerosol sizing. The size range is 10 - 0.3 μm.

Light Blocking (e. g. , Hiac Counter)

The light removed from a light beam due to the presence of a particle is proportional to the cross-section area of the particle. This statement is exactly true for large absorbing particles but is only approximately true for particles that scatter the light appreciably. White light is used rather than laser light to compensate for this and the instrument is used down to 1 μm. The size range is 100 - 1 μm.

SEDIMENTATION

Sedimentation has been widely used for size analysis and the method essentially measures the settling velocity of the particle and relates it to a sphere settling with some velocity. Gravity sedimentation covers the sub-sieve range down to about 1 μm and includes many well-established

techniques including the Andreason pipette. A disadvantage of the earlier
instruments was that a fixed height of fall was used and so to measure
small particles a long time was required. Modern instruments now vary
the height of fall and the time required for analysis has been reduced from
12 h to less than 20 min which compares well with other methods. One
such instrument the "Sedigraph" claims to measure down to 0.1 μm. This
is very tempting but the manufacturer states that measurements below 1 μm
must be viewed with caution and are acceptable if only less than 10% by
weight of the distribution is below 1 μm.

Of more general use is centrifugal sedimentation. An instrument
marketed by Joyce Ltd. was first developed by I.C.I. Ltd. (Dyestuffs
Division). The range of applicability is 10 μm to 0.1 μm and so is attractive. But assaying the small quantity of solids with accuracy is not easy
and depends very much on the material being measured. The suspension
can sometimes become unstable due to a density distribution. The settling
suspension must always be carefully observed with the strobescope provided. Gravity sedimentation size range 76-2 μm (0.1 μm under special
circumstances); centrifugal sedimentation size range 10-0.1 μm.

CALIBRATION MATERIALS

One important recent development to overcome the problems of
making accurate particle-size analysis is the production of well-documented
calibration materials. These have been produced under the aegis of
Bureau of Community Reference [6] and are available in the range 76-0.5
μm. These powders which have been riffled to obtain a large number of
samples which are as nearly identical as possible. These have been
analyzed by a number of specialist laboratories and the size distribution
certificated. They are sold complete with dispersion information so that
the analysis can be repeated as a test of an instrument or of technique.
Unfortunately these calibration materials are not yet available for the sub-
micron region but one covers the range 0.5-2.5 μm.

CONCLUSIONS

This paper has explained the reasons why size distributions measured
by different methods of analysis do not agree and in fact will only agree if
the particles are spherical. Even if the same instrument or method of
analysis is used, there may still be differences due to dispersion, sample
sub-division, or intrinsic error due to the finite number of particles
measured. With care, good agreement can be obtained but it is important
to maintain strict control of all conditions.

A brief review of the methods available has listed the methods available with comment as to their applicability to the range of particle size to be expected in coating systems.

REFERENCES

1. Lloyd, P. J. "Particle Size Analysis 1981," London: Heyden & Sons, 1981.
2. Harfield, J. G. and Knight P. "Particle Size Analysis 1981," London: Heyden & Sons, 1981.
3. Lloyd, P. J., Stenhouse, J. I. T. and Buxton, R. E. "Particle Size Analysis," London: Heyden & Sons, 1978.
4. Bernhardt, C. "Flussigkeiten und Zusatzsroff fur die Sedimentation Analyse," Freiburg; Forsch. Aufbereit, 1973.
5. British Standard 3406 Part 2, London: British Standards Institute, 1964.
6. Wilson, R. "Second European Symposium on Particle Characterisation," K. Leschonski, Ed., 1979.

BOOK

Particle Size Measurement, T. Allen, London: Chapman & Hall, 1979.

THE EFFECT OF PARTICLE-SIZE DISTRIBUTION
ON THE RHEOLOGY AND FILM FORMATION
OF LATEX COATINGS

Kenneth L. Hoy

Union Carbide Corporation
South Charleston, West Virginia

INTRODUCTION

The manner in which solid particles are naturally arranged in an element of volume is of extreme importance to the understanding of many physical properties and processes that we must deal with. In crystalline materials, the spacing of the molecules or atoms is geometrically regular; in liquids the arrangement is nonspecific and the spacings must be considered as a distribution of different geometric patterns that are best treated by statistical methods. The understanding of many liquid properties relies heavily on the development of lattice theory and imperfections or holes in the lattice spacing. At the other end of the spectrum of interest are the extremely complex interactions we experience in composites. A coating is an extremely complex composite which in a liquid form is applicable to a surface, and subsequently is changed by a film forming process to a solid composite film, in which pigments and fillers are dispersed in an amorphous glasslike polymer. The coating is expected to have a number of properties many of which are directly related to the arrangement of the particulate material (pigment and fillers) in the element of volume (vehicle).

The concept of critical pigment volume concentration (CPVC), first introduced by Asbeck and Van Loo [1], has been used to explain many of the performance properties of coatings. It is not the intent of the present

paper to reiterate the fundamental importance of this concept, but rather
to build a basic understanding of the effect that particle size and distributions have on packing arrangements and introduce a method of mathemati-
cally treating such relationships. In order to accomplish this goal some
basic definitions will be defined for the purpose of the ensuing discussion.

FUNDAMENTAL VOLUME RELATIONSHIPS

Consider an element of volume, V_e, in which there is dispersed or
suspended particulate matter of such size that the volume, a, of any individual particle is very small with respect to the volume of the element.
The total volume, V_i, of the particles is the summation of the individual
particles, and the ratio of the volume of particles to the element of volume,
V_e, which contains them is defined as the volume concentration, Φ.

$$V_i = \Sigma a_i \qquad (1)$$

$$\Phi = V_i/V_e \qquad (2)$$

If we continue to disperse particulate matter in the element of volume
until each particle is in contact with its nearest neighbors and no more can
be added, i.e., the element of volume is filled, then the ratio of the actual
volume of particles, V_c, contained in the element of volume to the volume
of the element, which is now the occupied volume, V_o, is called the
critical packing constant, Φ_c.

$$\Phi_c = V_c/V_o \qquad (3)$$

It is apparent that Φ_c is the fraction of the available volume that can be
occupied by the particles and the void fraction or porosity is given in $(1 - \Phi_c)$.

In the original dispersion, the volume concentration was less than
critical concentration Φ_c and the particles had ample space or "free
volume" to move about. The "free volume" of the system is estimated in
the following manner. The condition described may be modeled in Figure
7.1, in which the particles are dispersed in the volume, V_e. Suppose now
we could cause the particles to accumulate in one end of the element of
volume in a nonagglomerating way (Figure 7.2) and yet each particle in
contact with its nearest neighbors, the volume V_e is now the sum of the
free volume V_f and volume V_o occupied by the particles.

$$V_e = V_f + V_o \qquad (4)$$

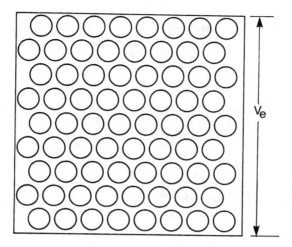

FIGURE 7.1   Model of a disperions of spherical particles in an element of
volume (V_e).

Since the volume occupied by the particles is given by $V_o = V_c/\Phi_c$ substitution into equation (4), dividing both sides of the equation by V_e, and
rearranging gives for the fractional free volume f of the system.

$$f = 1 - (\Phi/\Phi_c) \tag{5}$$

The fractional free volume is the key to understanding many of the physical
and rheological properties of coating materials. Indeed, nearly all
modern rheological equations relate the viscosity (η) and/or modulus to the
free volume of the system (Table 7.1). In the discussion thus far, no
attempt has been made to account for the interfacial region. Indeed, it is
treated as if it occupied no volume but rather is a surface with no thickness. In real systems, the interfacial region occupies a considerable
volume and a correction factor, α must be included. If a particle of volume (a) is surrounded by an interfacial layer of thickness X, then the volume
of the latter is S_aX where S_a is the surface area of the particle (Figure
7.3). The factor α is the ratio of the volume including the layer to the
volume of the particle without the layer.

$$\alpha = (a + S_aX)/a \tag{6}$$

If the particles are spherical then α is the ratio of the cube of the diameter
(D + 2X) of the particle with the absorbed layer to that (D) of the particle
without the layer (Figure 7.4).

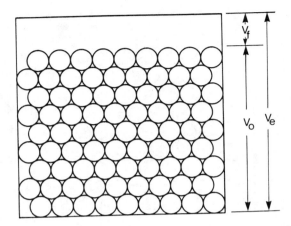

FIGURE 7.2   Model of dispersion of spherical particles illustrating the concept of free volume (V_f).

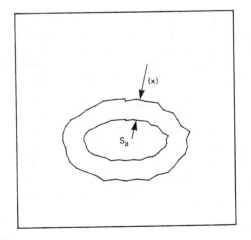

FIGURE 7.3   Correction factor for non-spherical particle [equation (6)].

TABLE 7.1
Free-Volume Relationship for the
Rheology of Dispersions

	Comments
Doolittle	
$\eta = A \exp(B/f)$	Suitable for pure liquids especially spherical molecules
Mooney	
$\eta = \eta_o \exp(2.5\Phi/f)$	Suitable for steady state shear for dispersions up to approximately 45% volume solids
Asbeck – Casson [18]	
$\eta^{1/2} = \eta_o^{1/2} f^a + K \dot{\gamma}^{-1/2} (f^a - 1)$	Relates shear rate $(\dot{\gamma})$ to viscosity
where $a = -A/2$ $K = B/A$	

$$\alpha = \left[\frac{D + 2X}{D} \right]^3 \tag{7}$$

Bassett and Hoy [2] found that the thickness of the hydration layer in certain acrylic latexes can be as large as 400-1000 Å. The absolute extent of the layer is a function of pH, ionic strength, acid content, etc., of the polymer. Figure 7.5 graphically portrays the value of the factor α when the layer is varied from 0 to 1000 Å and attached to a particle of 0.1 to 0.6 μm diameter. It is of interest to note that a layer of 400 Å on a 0.5 μm particle increases the volume of the particle by 40.5% over what was calculated from the hard-sphere model. Since materials absorbed or associated at an interface are known to have viscosities several orders of magnitude greater (10^2 to 10^6) than that of bulk liquid [3], it should be expected that much, if not all, of this layer contributes to the rheological properties of the dispersion by acting as though it were part of the particle itself. From these considerations, the free volume of the system must be corrected by the inclusion of the correction factor in equation (5) to give

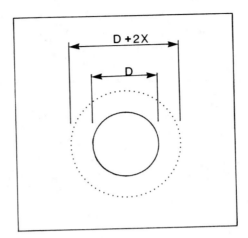

FIGURE 7.4   Correction factor (α) for spheres [equation (7)].

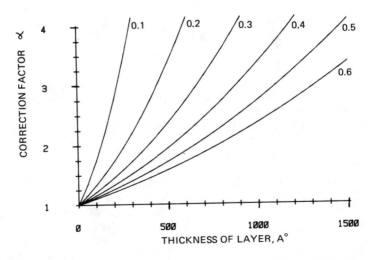

FIGURE 7.5   Correction factor (α) as a function of layer thickness for various particle sizes (μm).

equation (8). The value of Φ_c is generally regarded as independent of particle size, but in real distributions the increase in diameter will need to be considered.

$$f = 1 - (\alpha \, \Phi/\Phi_c) \tag{8}$$

Another method of correcting the volume concentration is by equation (9).

$$\Phi_e = \Phi_s + \Phi_p \tag{9}$$

Φ_e is the effective volume concentration, Φ_s is the volume of its interfacial material, and Φ_p is that of the particles. The relationship between Φ_s and α is given by

$$\Phi_s = \Phi_p (\alpha - 1) \tag{10}$$

Using this means of correcting, the free volume becomes

$$f = 1 - \frac{\Phi_s + \Phi_p}{\Phi_o} \tag{11}$$

The corrected free volume is the key factor necessary for the understanding of the rheology of dispersions, and basic to this understanding is a reliable means of estimating the packing characteristics of particles as they exist in the distribution.

It is the problem of packing of distributions that the remainder of this paper will consider. The approach will be to first review the packing character of uniform model particles, i.e., spheres, ideal binary packing, nonideal binary packing, and finally distributions.

PACKING OF UNIFORM SPHERES

Hard, uniform spheres can ideally be packed in at least four regular geometric arrangements. Each arrangement has a characteristic packing constant, and the number of nearest neighbors or coordination number is unique. Table 7.2 lists the four regular packings, their packing constant, Φ_c, and the coordination of each arrangement. Real systems consist of innumerable particles and, as such, seldom exist as a single arrangement throughout their volume. Rather, it is probable that the various arrangements exist side-by-side in micro-regions of the system. One can visualize that the actual packing is a distribution of these more regular packings. The random packing of spheres has been determined by a number of workers [4-7] in very diverse fields using very large numbers of uniform

TABLE 7.2
Regular Geometric Packing Arrangement
for Uniform Spheres[a]

Geometrical Arrangement	Packing Constant	Coordination Number
Cubic	0.5236	6
Orthothombic	0.6046	8
Tetragonal spheroidal	0.6981	10
Rhombohedral	0.7405	12

[a]Rarouki and Winterkorn, Highway Research Record No. 52,
10 (1964)

spheres. It is generally recognized that there are two different random
statistical arrangements of uniform spheres; dense-random and loose-
random packing. Table 7.3 lists a number of values from various investi-
gators. The results of Scott [5] and Bernal and Finney [6] are probably
the more reliable since these workers took precautions to eliminate wall
effects.

Dense-random packing is the more stable of the two arrangements
and occurs when a force, e.g., from an electronic double layer, is allowed
to act over a period of time. Kreiger and Hiltner [9] have described
latexes which in standing form iridescent colors characteristic of their
particle size. The latexes are carefully prepared with a minimum of
electrolyte and have very uniform particle sizes. They suggest that the
effect is caused by the repulsive force generated by the electronic double
layer which subsequently forces the particles into the arrangement which
has the greatest free volume. These authors suggest the hexagonal close-
pack configuration ($\Phi = 0.7405$), although it is not clear if the repulsive
force in conjunction with the induced surface orientation of the cell wall did
not create the pentagonal dipyramid structure described by Bagley [10]
which has a $\Phi_c = 0.72457$. Regardless of the exact structure it is apparent
that repulsive forces are sufficient to force most stable colloidal systems
to assume the maximum free volume or lowest energy state. Since most
dispersions contain catalyst fragments, ionic material, etc., it is suggested here that for practical systems the dense-random packing is near
what really exists. Rutgers [11] suggests from model experiments that
the packing of spheres in flowing suspension is different from those at
rest. From simulated colloidal suspensions in Couette flow he suggests

TABLE 7.3
Experimental Determination of Random
Packing of Uniform Spheres

Dense Random	Loose Random	Reference
0.629	—	4
0.637	0.601	5
0.634	0.601	6
0.657	—	7
0.602	—	8

that the packing characteristics are higher when flowing than at rest.
While his model does not account for repulsive forces, they do suggest
that a shear field can induce a more favorable packing arrangement.

Of particular interest is coordination number of a random packing
array. Bernal and Mason [12] determined from a unique experiment that
the coordination of dense-random packing is close to 8.5. Their experiment was based on packing uniform spheres in a rubber balloon, filling the
void space with paint, and then allowing the paint to drain and dry. Upon
disassembling the array each sphere was marked where it touched another
sphere. The number of contacts on each sphere was noted. Figure 7.6 is
a histogram based on the data reported by the authors. In an elegant
computer simulation of random packing, Levine and Chernick [13] estimated that coordination of dense-random packing is 8.52, although their
calculated packing density is only $\Phi_c = 0.601$, which is lower than might
be expected.

Since each packing arrangement has a unique coordination number,
empirical relationships have been developed to translate packing constant
Φ_c to coordination number. Smith et al. [14] developed the relation

$$N = 26.458 - 10.7262/\Phi_c \tag{12}$$

This particular relationship is not very accurate. Two other relationships
which have been found by the author to approximate more closely to actual
packing are

$$\Phi_c = 0.2210\sqrt{N} - 0.0160 \tag{13}$$

$$\Phi_c = N/(6.8929 + 0.7682N) \tag{14}$$

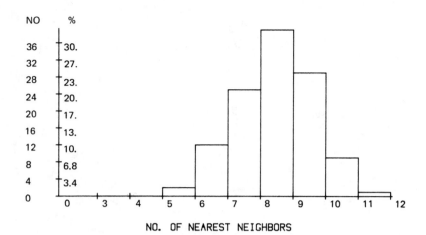

FIGURE 7.6   Coordination of randomly packed uniform spheres after
             Bernal and Mason [12].

Both of these empirical relationships fit the now regular packing arrangements to within 2.18% and have correlation coefficients of over 0.98.
Table 7.4 illustrates the accuracy of the relationships.

        Calculation of the coordination number of a random packed array of
spheres (0.635) from equations (12), (13), and (14) gives 9.56, 8.67, and
8.54, respectively.   The latter two compare favorably with the results of
Bernal and Mason [12].

                        IDEAL BINARY PACKING

        Consider again the element of volume, V_e, filled with uniform
particles of size $\underline{j}$; the critical packing constant Φ_j of the particles is the
ratio of the volume of the particles V_j to the volume of the element V_o.

$$\Phi_j = V_j/V_o \tag{15}$$

Suppose now that a smaller particle of size $\underline{i}$ can be placed between the
larger particles without causing a separation of the larger particles.   The
critical packing of the mixture Φ_{mix} is given by equation (16).

$$\Phi_{mix} = (V_j + V_i)/V_o \tag{16}$$

Since the volume of the element is V_j/Φ_j and $V_j/(V_j + V_i)$ is the volume
fraction F_j of the particles, substitution and rearrangement reduces equation (16) to equation (17).

TABLE 7.4
Comparison of Equations Relating
Coordination Number to Packing

N	Φ_c (Actual)	Φ_c (calc.) % error (equation (14))		Φ_c (calc.) % error (equation (12))		Φ_c (calc.) % error (equation (13))	
6	.5236	.5216	-0.38	.5236	0.00	.5253	+0.32
8	.6046	.6136	+1.49	.5802	-4.04	.6091	0.74
10	.6981	.6861	-1.72	.6506	-6.80	.6829	-2.18
12	.7405	.7448	+0.58	.7405	0.00	.7496	+1.22

$$\Phi_{mix} = \Phi_j / F_j \tag{17}$$

Now as we continue to place particles of size $\underline{i}$ in the element of volume, all of the void space between the larger particles $V_0 - V_j$ will be filled with the smaller particles. The packing is still given by equation (15). However, since V_i is equal to the packing constant of the smaller particles Φ_i times the void space between the particles $1-\Phi_j$, equation (15) reduces to equation (18). The volume fraction of larger particles at the maximum packing is obtained by equating (17) and (18) and solving for F_j (at max) giving equation (19).

$$\Phi_{max} = \Phi_j + \Phi_i (1 - \Phi_j) \tag{18}$$

$$F_j \text{ (at max)} = \frac{\Phi_j}{\Phi_j + \Phi_j (1-\Phi_j)} = \frac{\Phi_j}{\Phi_{max}} \tag{19}$$

By a similar reasoning the volume relationship that exists when there are more smaller particles than are required to fill the void space, is given by equation (20).

$$\Phi_{mix} = \frac{\Phi_i}{1 - F_i (1-\Phi_i)} \tag{20}$$

Φ_i is the packing constant and F_i is the volume fraction of the smaller particles, respectively. Equations (17), (18), (19), and (20) define the packing characteristics of all compositions of particles of sizes $\underline{j}$ and $\underline{i}$.

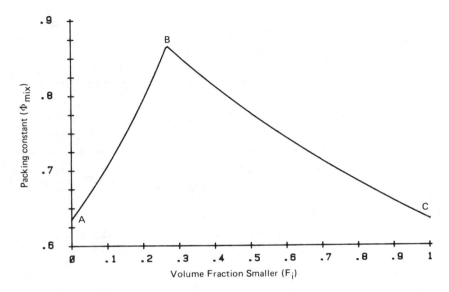

FIGURE 7.7   Ideal packing for binary spheres; A-B, Case I; Case II, B; Case III, B-C.

Figure 7 represents the various kinds of packing that can exist; it is in a sense a type of phase diagram. Equation (17) applies when there is a preponderance of larger particles, Case I, but below the most dense packing. Case II, the most dense packing, is given by equations (18) and (19). Equation (20) applies when there are more small particles than will conveniently fit between the larger particles, Case III.[*]

NONIDEAL BINARY PACKING

In real systems, we cannot expect that small particles can exactly fit between the larger particles without causing dislocations in the packing lattice of the larger particles. The major causes for the failure of the equations derived for ideal packing are: (1) separation of larger particles by smaller particles; (2) boundary effects between the micro regions of the binary packing, i.e., the interface of larger particles with small particles; and (3) inability of smaller particles to fit exactly in the void space of the larger particles.

---

[*]See acknowledgment.

The problem may be experimentally treated if we assume that a single correction factor, packing efficiency, can be evaluated as a function of the relative sizes of the particles. A convenient function is the ratio of the small particle diameter to the larger diameter

$$R = D_{sm}/D_{lg} \tag{21}$$

The efficiency factor, E, as conceived here would vary between the limits of 1.00, when the value of $\underline{R}$ approaches 0, and 0, when the value of $\underline{R}$ approaches 1. Thus, when the diameter of the smaller particle is equal to the larger, no increase in binary packing would be evident, and when the diameter of the larger particle is very large with respect to the smaller particle, the binary packing approaches the ideal packing equation derived previously. If we approximate the efficiency factor by a quadratic function of the type

$$E = c + bR + aR^2 \tag{22}$$

then it can be shown that the equations derived previously can be modified to accurately predict the packing-volume relationships for binary mixing. Again, it is convenient to treat the system as three distinct sets of equations.

Case I

The packing factor Φ_{mix} is defined as

$$\Phi_{mix} = \frac{V_j + V_i}{V_o} \tag{23}$$

Earlier it was assumed that the small particles always fit between the larger and the occupied volume, V_o, was equal to the volume occupied by the larger particles. In a real system, the smaller particles cause an increase in the overall occupied volume, and hence

$$V_o = V_j + V_i (1 - E_1) \tag{24}$$

where the term $V_i (1 - E_1)$ accounts for the inefficiency of packing. Substituting (24) into (23) and making the usual rearrangement yields

$$\Phi_{mix} = \frac{\Phi_j \Phi_i}{F_j \Phi_i + F_i (1 - E_1) \Phi_j} \tag{25}$$

Equation (25) reduces to the ideal packing when $E = 1$.

Case II

Maximum packing occurs when the void volume is filled with the smaller particles and because the volume of the smaller particles required is less than that expected. Ideally we can write

$$V_i = E_2 \Phi_i (V_o - V_j) \tag{26}$$

in which the volume required to fill the void space is reduced by the factor E_2. The maximum packing constant now becomes

$$\Phi_{max} = \Phi_j + \Phi_i (1 - \Phi_j) E_2 \tag{27}$$

Equating equations (25) and (27) under Case II conditions and solving for F_i gives equation (28).

$$F_{i_{max}} = \frac{\Phi_i \Phi_j}{[\Phi_j + \Phi_i (1-\Phi_j) E_2][\Phi_i - (1-E_1) \Phi_j]} - \frac{\Phi_i}{[\Phi_i - (1-E_1) \Phi_j]} \tag{28}$$

The value of E_1 will be shown later to be equal to 1 for spherical particles and equation (28) reduces to

$$F_i = 1 - \frac{\Phi_j}{[\Phi_j + \Phi_i (1 - \Phi_j) E_2]} \tag{29}$$

In the case of ideal packing, E_2 and E_1 are by definition equal to 1 and equation (29) again reduces to Case II ideal packing.

Case III

When conditions exist in which there are more smaller particles than required to fill the voids between the large

$$V_e = V_o^j + V_o^i - (V_o - V_j) E_3 \tag{30}$$

Performing the usual substitutions and rearrangements gives equation (31).

$$\Phi_{mix} = \frac{\Phi_i \, \Phi_j}{\Phi_i \, F_j \, (1 - E_3) + \Phi_j \, (1 - F_j) + \Phi_i \, \Phi_j \, E_3 \, F_j} \qquad (31)$$

Equation (31) reduces to the simpler ideal packing equation when $E_3 = 1$.

The packing equations derived are quite general and can be applicable to any geometrical shape. However, the actual values of Φ_j, Φ_i, and E_1, E_2 and E_3 must be evaluated independently and may be variable within a process. Thus, one can visualize a process where the alignment of cylindrical particles depends on type of stresses applied. However, in the case where the particles are spherical or can be approximated by spheres, then the equation will reduce to the following forms:

Case I

$$\Phi_{mix} = \frac{\Phi}{1 - E_1 \, F_s} \qquad (32)$$

Case II

$$\Phi_{max} = \Phi + E_2 \, (1 - \Phi) \qquad (33)$$

$$F_s = \frac{E_2 \, (1 - \Phi)}{1 + E_2 \, (1 - \Phi)} \qquad (34)$$

Case III

$$\Phi_{mix} = \frac{\Phi}{1 - F_j \, E_3 \, (1 - \Phi)} \qquad (35)$$

When the particles are spherical the different efficiency factors, namely E_1, E_2, and E_3, can be evaluated with respect to each other by equating both Case I equations to Case III at the maximum packing condition of Case II. Such an evaluation is possible when the particles are spherical. From this analysis it is apparent that $E_1 = 1$ and $E_2 = E_3$.

The data of Yerazunis et al. [15] provide a convenient means to evaluate the emperical constants of equation (22). Table 7.5 summarizes the data. It is to be noted that all of the Yerazunis data is in the Case III area; if equation (35) is solved for E, then it is possible to calculate the efficiency.

TABLE 7.5
Experimental Data of Yerazunis et al. [15]

| Ratio Sm/Lg | Packing constant at volume fraction large | | | |
	0.4	0.5	0.6	0.7
.3876	.6896	.7008	.7190	.7131
.2933	.7014	.7174	.7337	.7483
.1742	.7146	.7364	.7611	.7855
.0833	.7275	.7632	.7816	.8135
.0148	.7400	.7710	.8049	.8407
.0056	.7428	.7751	.8095	.8477
.000+	.7467	.7797	.8157	.8534

$$E = \frac{(\Phi_{mix} - \Phi_{sphere})}{\Phi_{mix} F_j (1 - \Phi_{sphere})} \tag{36}$$

Table 7.6 summarizes these calculations for various values of Sm/Lg. Regression analysis of the data in Table 7.6 is now a convenient means of evaluating the constants in equation (22). The results are quite gratifying; Figure 7.8 is a plot of the data. The correlation coefficient is 0.9966. The equation is

$$E = 0.9832 - 1.458R + 0.4758R^2 \tag{37}$$

Using this equation to evaluate the efficiency factor, it is now possible to test the nonideal packing equations (25), (27), and (31). Figure 7.9 illustrates the theoretical packing of binary spheres over a range of R(Sm/Lg = 0 to 1). From these equations it is expected that the amount of smaller material required to reach maximum packing decreases as the ratio approaches 1. This is in contrast to the analytical expressions developed by Lee [16], who treated the amount of smaller particles required to reach the maximum as a system constant.

Interestingly, if the data point off the curve in Figure 7.8 is omitted from the regression, the correlation coefficient is improved from 0.9966 to 0.9984. Of greater interest is the fact that the regression coefficients suggest the following empirical relations:

$$E = 1 - (2 - \Phi_i \Phi_j) R + \Phi_i \Phi_j R^2 \tag{38}$$

TABLE 7.6
Efficiency Factor for Spheres[a]

Ratio Sm/Lg	Packing constant at volume fraction large			
	0.4	0.5	0.6	0.7
.3876	.5185	.4964	.5197	.4165
.2933	.6259	.6127	.6015	.5824
.1742	.7418	.7393	.7455	.7416
.0833	.8501	.8460	.8466	.8519
.0148	.9533	.9540	.9553	.9519
.0056	.9757	.9781	.9760	.9766
< .001	1.0066	1.0049	1.0036	.9964

[a]Calculated from the Case III nonideal packing equation using the data from
Yerzunis et al. [15].

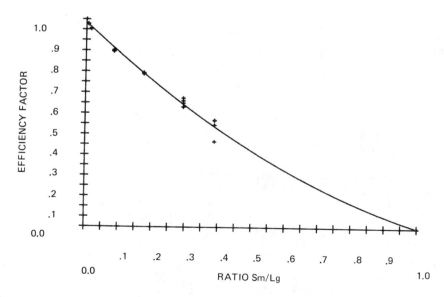

FIGURE 7.8   Plot of efficiency factor vs. ratio of radii; (+) data points,
Yerazunis et al. [15]; solid line is equation (37).

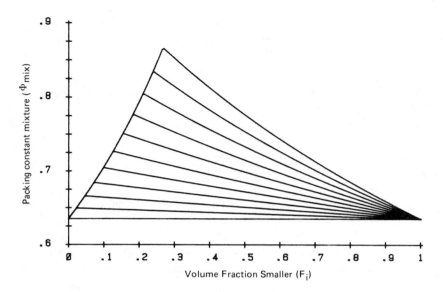

FIGURE 7.9   Binary nonideal packing of spheres; ratio of smaller/larger
varying from 0 to 1.0.

$$E = 1 - \sqrt{2}\ R + (\sqrt{2} - 1) R^2 \tag{39}$$

Equation (38) reduces to (40) if the particles are both spherical.

$$E = 1 - (2 - \Phi^2) R + \Phi^2 R^2 \tag{40}$$

A test of the equation (39) is afforded by the data of Yerzunis et al. [14].
Using equation (39) it is possible to calculate the packing constant for
spheres.   Table 7.7 gives the calculated value of the packing constant for
uniform spheres.   The average value is 0.6346 ± 0.005.

A test of the general equation (38) is provided by data from a model
system studied by the author several years ago.   The model consisted of
6 mm glass beads ($\Phi_c = 0.629$) and common table salt ($\Phi_c = 0.573$) contained in a 500 ml graduated cylinder.   Calculation of maximum packing
and volume fraction of small particles (salt) gives: $\Phi(\text{max}) = 0.812$ (calc.),
0.813 (found) and $F_i = 0.225$ (calc.), 0.239 (found).

TABLE 7.7
Calculation of Packing Constant for Spheres:
Data Yerzunis et al. [15]

Ratio Sm/Lg	Φ calc.	E calc.	F_1	Φ_{mix}
.3876	.6383	.5141	.4000	.6896
.2933	.6384	.6208	.4000	.7014
.1742	.6346	.7662	.4000	.7146
.0833	.6330	.8851	.4000	.7275
.0148	.6339	.9792	.4000	.7400
.0056	.6353	.9921	.4000	.7428
.3876	.6351	.5141	.5000	.7008
.2933	.6364	.6208	.5000	.7174
.1742	.6328	.7662	.5000	.7364
.0833	.6298	.8851	.5000	.7532
.0148	.6321	.9792	.5000	.7710
.0560	.6500	.9921	.5000	.7751
.3876	.6389	.5141	.6000	.7190
.2933	.6335	.6208	.6000	.7337
.1742	.6325	.7662	.6000	.7611
.0833	.6266	.8851	.6000	.7816
.0148	.6299	.9792	.6000	.8049
.0056	.6323	.9921	.6000	.8095

$\Phi(avg)$ = 0.63464 ± 0.005
Efficiency factor calculated $E = 1 = \sqrt{2}\ R + (\sqrt{2} - 1)\ R^2$

PACKING CHARACTERISTICS OF DISTRIBUTIONS

There are occasions when it is possible to treat a system as a
binary mixture, e.g., concrete is a mixture of sand and gravel, but in
real systems we must most often deal with a distribution of particles. The
packing characteristics of a distribution must be a function of the nature
of the distribution and the breadth of the distribution. If these elements
are known then it is possible to estimate the packing of that system. To
illustrate the method we will assume that the frequency of particles
follows a Gaussian distribution (it is much easier to deal with a volume
distribution), the particles are spherical and that the standard deviation

from the mean is known. Further, it will be assumed the distribution can
be approximated by a histogram of discrete size intervals.

The method proposed is to treat the two largest intervals as a
binary mixture; calculate its binary packing constant (Φ 1, 2) from the
previously derived equations. The combined fractions are now considered
to be a unique packing array (Φ 1, 2) in which the next smaller interval
must be interspersed to give a new packing array (Φ 1, 2, 3). The process is repeated until all of the intervals have been incorporated. Figure
7.10 illustrates the mathematical process. The method is ideally suited
to the high-speed digital computer. Care must be taken to avoid creating
more intervals than are required to adequately describe the distribution,
otherwise roundoff error will cause the relationships to diverge. The
calculation was performed over the range of $\pm$ 2 σ or 95% of the distribution.

Figure 7.11 is a plot of the calculated packing of a normal distribution which is centered about 1 μm with a standard deviation (σ) of
$\pm$ 0.00001 to 0.99. The implication of Figure 7.11 is that the broadening
of the distribution increases the packing and that in this case the limit of
the packing approaches the asymptotic value of ~0.69. The method can
be used to calculate the packing of any type of distribution provided a reasonable approximation of the particle size distribution can be made.

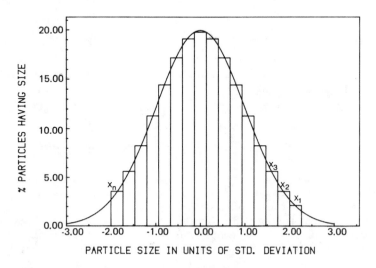

FIGURE 7.10   Calculation of packing constant of a distribution using
binary packing relationships.

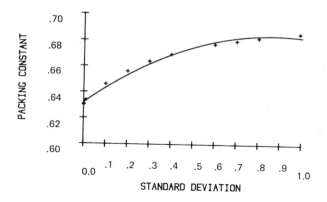

FIGURE 7.11   Calculated packing (Φ) of a Gaussian distribution of
spherical particles centered about 1 μm as a function of
its standard deviation (σ).

PARTICLE PACKING AND LATEX RHEOLOGY
AND FILM FORMATION

The relationship that particle packing has on the rheological properties of latexes is apparent from the free volume of the system. The
greater the tendency of the particles to pack efficiently the more free
volume will be available and hence the lower the viscosity. The Mooney
relationship [17] when modified by packing and layer considerations is able
to predict the rheology up to 50% - 60% volume concentration. The findings of Rutgers [10] indicate that shear fields can induce more favorable
packing arrangements, and thus the increasing dynamic free volume provides a ready explanation of the pseudo-plastic flow (shear-thinning). However, experience tells us that the effect of a static boundary as opposed
to a moving boundary (flow) can alter the packing as well.

The effects that are cuased by boundaries are quite pronounced in
coatings. A coating is, in practice, a thin layer of material between two
boundaries, the substrate, and the air-liquid interface. While it is apparent that boundaries under certain conditions can induce favorable packing
arrangements, the advent of two boundaries in close proximity only causes
severe packing constraints. Thus, if we assume that random packing
exists in a system (coordination number 8.5), then at the interface it is not
possible for the nearest neighbors normally present to exist. The net
result is that the coordination is decreased to 6-7. From equation (13) we
can calculate that the packing for uniform spheres would be reduced from
0.635 to 0.57. This dramatic reduction of packing causes the viscosity of
the boundary layer to be much higher than the bulk. An even more

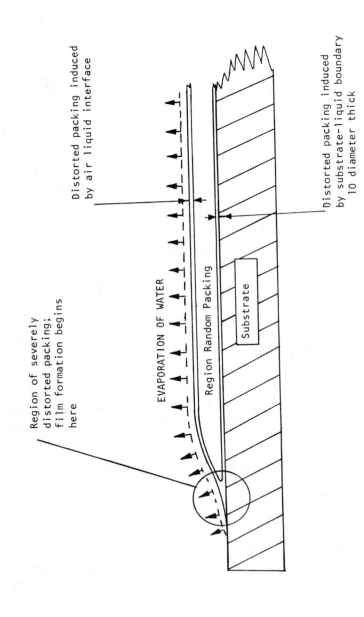

FIGURE 7.12 Model of latex film formation.

pronounced effect is observed when two boundaries, such as at an edge, converge. The edge must further constrain the packing even more, perhaps another 2-3 nearest neighbors are not allowed, and the coordination is further reduced from 6-7 to 4.5-5.5. This would mean that the packing is now near 0.49. The effect is to raise the viscosity of the edge boundary to many orders of magnitude above the bulk. Because packing is so constrained, the slightest loss of liquid through evaporation means that it is the edge and thin places that reach critical packing first. At critical packing the further loss of any more liquid through evaporation causes the surface forces to exert amazingly high compressional forces on the particles; they are literally forced through the repulsive layers and coalescence takes place. Hence, the film-forming process always begins at the edge and moves toward the middle of the coated surface.

Figure 7.12 pictures this model of film formation. The same boundary constraints apply to both solvent borne and latex paints. However, the free volume is even more constrained in the latex system because the vehicle itself is in particulate form. Only after film formation are the pigments and fillers incorporated into the binder. On this basis the theory explains very well the application properties of quick set, open time, lap characteristics, flow, and leveling when compared to solvent-borne paints. One added bit of evidence in support of the theory that everyone has observed is when paint is accidently dropped. During clean-up, a ring has developed around the edge indicating that film formation has begun at the edge of the film.

ACKNOWLEDGMENT

The author would like to dedicate this work to the memory of the late Dr. E. A. Brecht, Dean of the School of Pharmacy of the University of North Carolina, who encouraged the author to develop the original Ideal Packing Theory during his tenure as instructor in Pharmacy in 1954 and 1955. The equations have been restructured and modified to apply to colloidal and latex systems.

REFERENCES

1. Asbeck, W. K. and Van Loo, M. Ind. Eng. Chem., 41:1470, 1949.
2. Bassett, D. R. and Hoy, K. L. "The Expansion Characteristic of Carboxylic Polymers," Polymer Colloids, II, R. M. Fitch, Ed.; New York: Plenum Publishing Corporation, 1981.
3. Criddle, D. W. "Viscosity and Elasticity of Interfaces," Rheology, Vol. 3, F. R. Eirich, ed. New York: Academic Press, 1960.
4. Rice, O. K. Chem. Phys. 12: 1, 1944.

5.   Scott, G. D. Nature 188:908, 1960.

6.   Bernal, J. D. and Finney, J. L. Nature 214:265, 1967.

7.   Susskind, H. and Becker, W. Nature 212:1564, 1966.

8.   Westman, A. E. R. and Hugill, H. R. J. Am. Ceram. Soc. 13:767, 1930.

9.   Kreiger, I. M. and Hiltner, P. A. J. Phys. Chem. 73:2386, 1969.

10. Bagley, B. G. Nature 208:674, 1965.

11. Rutgers, R. Nature 143:465, 1962.

12. Bernal, J. D. and Mason, J. Nature 188:910, 1960.

13. Levine, M. M. and Chernick, J. Nature 208:68, 1965.

14. Smith, W. D., Foote, P. D. and Busang, P. F. Phys. Ref. 36:524, 1930.

15. Yerazunis, S., Cornell, S. W. and Winkner, C. Nature 207:835, 1965.

16. Lee, D. I. J. Paint Technol. 42:580, 1970.

17. Frisch, H. L. and Simha, R. "Viscosity of Colloidal Suspensions," Rheology, Vol. 1, F. R. Eirich, ed. New York: Academic Press, p. 596, 1956.

18. Asbeck, W. K. Off. Digest., 33:70, 1961.

CORRELATION BETWEEN COLOR STABILITY OF PAINT SYSTEMS AND PIGMENT—SOLVENT WETTABILITY

T. Satoh

Research Laboratory, Kansai Paint Co., Ltd.
Hiratsuka, Kanagawa, Japan

ABSTRACT

In order to clarify the details of interactions between pigments, resins, and solvents, correlations between adhesion-tension of pigment with solvent and solubility parameters of solvent were examined. A distinct correlation was found to exist between the hydrogen-bonding parameter (δ_H) of solvents and the adhesion tension of pigments with solvents. The δ_H value of solvents giving the maximum adhesion tension varies with each pigment. The dispersability of pigments and polyvinyl-chloride resin powders in vehicles was found to be dominated by the wettability. The color stability of paints on storage is discussed in connection with the same reasoning and results.

INTRODUCTION

It is well known that there exists a definite correlation between the color-stability of paints on storage and the dispersion stability of the pigments employed.

One of the biggest difficulties in studying the exact mechanism of the pigment dispersion, however, has been that no practical means enable us to fully clarify the details of the interaction between pigments, resins, and solvents in a quantitative way. As an approach to such method, the penetration-rate method has been applied to determine the wettability of powders with liquids. In recent papers [1-3], Tanaka and Koishi presented an improved penetration-rate measurement apparatus and discussed the analytical procedure for the results obtained.

In an attempt to obtain a more thorough understanding of related factors, we have examined the correlation between adhesion tension of pigments with organic solvents and Hansen's solubility parameter, and tried to relate the dispersability of pigments in vehicles to the color stability in storage.

EXPERIMENTAL

Materials

The ten commercial, organic and inorganic pigments are shown in Table 8.1, and the 15 solvents were of standard reagent grade and are tabulated in Table 8.2.

Measurement of Adhesion Tension

Apparatus. The block diagram of the apparatus used in our experiments is shown in Figure 8.1. The amount of the solvent penetrated into the sample powder bed was weighed directly with an electronic balance (Mettler PR-1200). The essential parts of apparatus were set up in the air cabinet saturated with the vapor of the solvent to be used and experiments were carried out in the constant temperature room at 20°C.

Procedure. The glass tube is shown in Figure 8.2. One end of the glass tube was closed up with filter paper. The sample tube was tapped 100 times with a tapping machine (Toyo Scientific Co., Ltd.). After inserting a glass rod into the glass tube, the sample powder bed was compressed with a universal tensile machine at 50 kg/cm^2 pressure for 20 min (Shinkoo Tsuushin TOM-500). About 500 ml of the solvent were poured into vessel which was attached to the slider. The glass tube packed with the sample powder was then set on the sample holder, and dipped into the vessel by operating the slider until the buoancy of the solvent reached 50 mg weight, and the weights of the penetrated solvent were then recorded on a recorder. One example is given in Figure 8.3.

Analytical procedure. Tanaka and Koishi [1] developed equations applicable for analyzing the penetration process of liquids into a powder bed, and discussed the relationship between chemical structure and immersional tension as well as adhesion tension.

TABLE 8.1
Properties of Pigments

Pigments	Producer	Physical properties: Specific gravity 25°C	Specific[a] surface area (m^2/g)	Mean particle[b] size (μm)	Particle[c] shape
Iron oxide	Toda industry	5.20	11.85	0.30	Granular
Titanium dioxide	Teikoku kakou	4.10	16.89	0.25	Sphere
Molybdate orange	Kikuchi shikiso	5.30	6.64	—	Granular
Insoluble azo. A	Hoechst	1.57	19.74	0.30	Flaky
Insoluble azo. B	Bayer	1.40	13.65	0.2–1.8	Flaky
Insoindolinone	BASF	1.70	61.69	—	Flaky (platty)
Quinacridone	Toyo souda	1.50	43.47	0.1~0.5	Flaky (platty)
Dioxazine	Sumitomo chemical	1.67	48.23	0.03~0.07	Flaky (platty)
Phthalocyanine blue	Toyo Ink Mfg.	1.70	39.54	—	Flaky
Carbon black	Mitubishi kasei	1.90	99.46	0.03	Sphere

[a]Measured by BET methods (N_2)

[b]Refer to reference No. 13

[c]Determined by electron micrographs

TABLE 8.2
Properties of Solvents

	Physical properties (at 20°C)			Hansen parameter			
	Density (g/cm^3)	Viscosity (C. P.)	Surface tension (dyne/cm^2)	δ	δ_d	δ_p	δ_H
Methyl alcohol	0.792	0.593	22.5	14.28	7.42	6.0	10.9
Ethyl alcohol	0.791	1.194	22.3	12.92	7.73	4.3	9.5
n-Propyl alcohol	0.805	2.256	23.7	11.97	7.75	3.3	8.5
n-Butyl alcohol	0.811	2.95	24.6	11.30	7.81	2.8	7.7
Propylene glycol	1.038	15.08	72	14.8	8.24	4.6	11.4
Ethylene glycol	1.112	5.56	—	16.3	8.25	5.4	12.7
Ethyl ether	0.720	0.377	17.0	7.62	7.05	1.4	2.5
Dioxane	1.034	1.31	35.4	10.00	9.30	0.9	3.6
Aniline	1.021	1.051	42.9	11.04	9.53	2.5	5.0
Dimethyl formamide	0.952	0.922	36.8	12.14	8.52	6.7	5.5
Benzene	0.881	0.694	28.9	9.15	8.95	0.5	1.0
Toluene	0.865	0.59	28.5	8.91	8.82	0.7	1.0
Xylene	0.865	0.664	28.9	8.80	8.65	0.5	1.5
Cyclohexane	0.777	0.771	24.9	8.18	8.18	0.0	0.0
Acetic acid	1.05	1.08	27.4	10.5	7.1	3.9	6.6

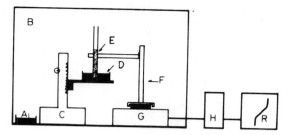

FIGURE 8.1   Apparatus for penetration rate measurements. B: air bath,
A: solvent vessel, C: up-down slider, G: top loading electronic balance, D: solvent vessel, E: glass tube packed with
samples, F: sample holder, H: suppressor, R: recorder.

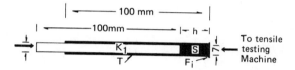

FIGURE 8.2   The glass tube packed by samples. S: sample, K_1: glass
rod, T: glass tube, F_i: filter paper.

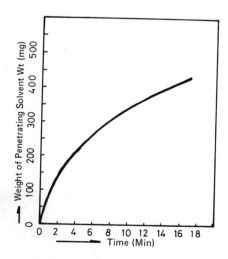

FIGURE 8.3   Weight changes of penetrating ethyl alcohol with time for
the molybdate orange.

By converting h_t (penetration distance) in Washburn's equation to W_t using $W_t = \epsilon \rho S h_t$ and considering the tortuosity ratio k, the rate of penetration is

$$U = dW_t/dt = K(W_\infty /W_t - 1) \tag{1}$$

where

$$K = \frac{\bar{\gamma}_\ell^2 \, \epsilon S \rho \, \ell^2 g}{8k^2 \eta} \tag{2}$$

At an early stage of penetration, we may assume $W_\infty /W_t \gg 1$, so that

$$dW_t/dt = KW_\infty /W_t \tag{3}$$

By integration, we obtain

$$W_t^2 = 2KW_\infty t \tag{4}$$

and at $t = 1$, $1/2 \, W_{t=1}^2 = KW_\infty$ $\tag{5}$

We calculated KW_∞ from equation (5), as the value $W_t = 1$ was obtainable from plotting $\log W_t$ vs. $\log t$ as shown in Figure 8.4.

At equilibrium in the penetration process ($t = \infty$),

$$\gamma_\ell \cos \theta = \frac{\bar{\gamma}_\ell g W_\infty}{2 \epsilon S} \tag{6}$$

From equations (2) and (6), we derive

$$\frac{4\eta (KW_\infty)}{\epsilon^2 S^2 \bar{\gamma}_\ell \rho_\ell^2} = \frac{\gamma_\ell \cos \theta}{k^2} \tag{7}$$

to define $\gamma_\ell \cos \theta /k^2$ as the adhesion parameter. All the parameters can be determined experimentally.

$$\frac{\eta}{\rho_\ell^2} (KW_\infty) = \frac{\epsilon^2 S^2 \bar{\gamma}_t \gamma_\ell \cos \theta}{4k^2} = A \tag{8}$$

Here we define A as the adhesion tension parameter which is obtainable from experimental data.

$\bar{r}_t$ was determined from equations (9) and (10).

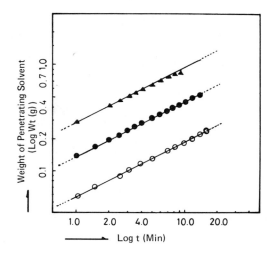

FIGURE 8.4   Weight changes of penetrating solvents with time for the molybdate orange. Ethyl ether (▲), ethyl alcohol (●), ethylene glycol (o).

$$\epsilon = \frac{V - W/\rho}{V} \qquad (9)$$

where $V = Sh$. Thus

$$\bar{r}_t = \frac{1}{Sw\rho} \cdot \frac{\epsilon}{1 - \epsilon} \qquad (10)$$

W_t: Mass of penetrating solvent at time t
W_∞:   Mass of penetrating solvent at equilibrium
ϵ: Porosity of powder bed
S: Cross-section area of powder bed
ρ_ℓ: Density of penetrating solvent
η: Viscosity of the solvent
g: Acceleration of gravity
Θ: Contact angle
V: Volume of sample bed
h: Length of sample bed
ρ: Density of sample powder
Sw: Specific surface area of sample powder
W: Mass of sample powder
$\bar{r}_\ell$:   Hydrodynamic mean capillary radius
$\bar{r}_t$:   Porosity mean capillary radius
A: Defined adhesion tension parameter

TABLE 8.3

Determinations of the Penetration Rate for the
Molybdate Orange with Solvents
By the Equation $W_t^n = 2KW_\infty t$

Solvent	Wt=1 (t=1 sec)	n
Methyl alcohol	2.33×10^{-2}	1.969
Ethyl alcohol	1.82	2.000
n-Propyl alcohol	1.37	1.908
n-Butyl alcohol	—	—
Propylene glycol	—	—
Ethylene glycol	0.77	2.000
Ethyl ether	3.78	1.938
Dioxane	2.88	2.000
Aniline	1.97	1.938
Dimethyl formamide	3.31	2.000
Benzene	3.79	2.000
Toluene	—	—
Xylene	—	—
Cyclohexane	2.37	1.969
Acetic acid	2.98	1.938

RESULTS AND DISCUSSION

Experimental Results

The packing properties of samples are summarized in Table 8.4. The adhesion tension parameter A of pigments with solvents is tabulated in Tables 8.5 and 8.6. Adhesion tension parameters of pigments with solvents are summarized in Tables 8.7 and 8.8. In this case, porosity mean capillary radius were used in place of hydrodynamic mean capillary radius in equation (7). Because the porosity mean capillary radius is not equal to the hydrodynamic radius in physical terms, the results summarized in Tables 8.7 and 8.8 are not absolute values but relative values. Values of adhesion tension parameter are distributed in the range from 24.2×10^{-3} g/cm^2 to 0.1×10^{-3} g/cm^2, that is from 23.7 dyne/cm^2 to 0.1 dyne/cm^2. If the tortuosity [4] ratio of the sample powder bed is simply 1.4, the values of adhesion tension are distributed in the range of 46.5 dyne/cm^2 to 0.2 dyne/cm^2.

TABLE 8.4
Packing Properties of the Sample

Pigments	Packing mass (g)	Packing heights h (cm)	Packing volume (cm^3)	Porosity ϵ	Mean capillary radius $r \times 10^6$	$1/4\,\epsilon^2 S^2 r \cdot g^2 \times 10^2$
Iron oxide	1.80	1.71	0.346	0.470	1.52	1.19
Titanium dioxide	1.50	2.54	0.366	0.619	2.35	3.19
Molybdate orange	1.80	2.60	0.339	0.660	5.52	8.52
Insoluble azo. A	0.55	1.54	0.350	0.393	2.09	1.15
Insoluble azo. B	0.50	1.91	0.357	0.597	3.53	4.47
Insoindoli-none	0.60	2.13	0.353	0.563	0.74	0.83
Quinacri-done	0.52	1.93	0.347	0.525	1.69	1.66
Dioxazine	0.60	2.52	0.359	0.627	2.09	2.92
Phthalo-cyanine blue	0.60	1.86	0.333	0.519	1.61	1.54
Carbon black	0.60	1.54	0.316	0.452	0.436	0.32

Correlation Between Adhesion Tension
and Hansen's Parameters

As shown in equation (6), adhesion tension is deominated by W_∞, ϵ and $\bar{r}_l$. It is very difficult to determine the adhesion tension precisely from these experiments because the precise determination of hydrodynamic mean capillary radius and flow character in a capillary is impossible.

But for samples which are prepared using the same powder and the same packing condition, the hydrodynamic capillary radius, its distribution

TABLE 8.5
Adhesion Tension Parameter A Estimated by the
Equation (8)
$(\times 10^{5}$ (g weight)$^2 \eta/\rho_{\ell}^{2})$

	Iron oxide	Titanium dioxide	Molybdate orange	Insoluble azo. A	Insoluble azo. B
Methyl alcohol	9.49	0.92~1.33	25.8	6.70	15.0
Ethyl alcohol	7.16	0.19	31.6	10.7	14.8
n-Propyl alcohol	8.70	1.74	32.5	5.32	19.5
n-Butyl alcohol	—	1.46	—	—	—
Propylene glycol	2.08	—	—	—	8.40
Ethylene glycol	1.64	—	13.5	4.44	35.1
Ethyl ether	1.97	—	51.8	11.7	11.3
Dioxane	1.61	29.8	62.7	6.27	22.2
Aniline	0.34	—	19.0	4.96	23.4
Dimethyl formamide	10.8	—	56.0	2.77	27.5
Benzene	10.8	31.6	64.1	11.0	12.3
Toluene	—	9.91	—	—	—
Xylene	—	37.7	—	—	—
Cyclohexane	5.21	24.9	36.0	5.49	16.1
Acetic acid	8.81	—	43.3	3.53	21.9

and hydrodynamic flow properties in a capillary must be same. If so, adhesion tensions of the pigments with various organic solvents compare with each other satisfactorily.

For this reason, we examined the correlation between adhesion tension of pigments with organic solvents and Hansen's [5] solubility parameter using the adhesion tension parameter A instead of adhesion tension.

Results are summarized in Figures 8.5, 8.6, 8.7 and 8.8 in which the adhesion tension parameter of pigment with organic solvents is plotted against Hansen's solubility parameter. From the figures it is quite clear that a distinct correlation exists between the hydrogen-bonding parameter (δ_H) value of solvents giving a maximum adhesion tension parameter A varies with the pigments as shown in Figure 8.9 (hereafter we call this relation between A and δH as the adhesion spectrum).

As a result of these experiments, the adhesion spectrum of pigments studied can be divided into the following three types:

TABLE 8.6
Adhesion Tension Parameter A Estimated by the
Equation (8)
$(\times 10^5 \text{ (g weight)}^2\, \eta/\rho_\ell^2)$

	Isoindoli-none	Quinacri-done	Dioxazine	Phthalo-cyanine blue	Carbon black
Methyl alcohol	6.75	8.75	9.59	8.83	4.25
Ethyl alcohol	13.7	3.07	0.87	7.16	5.99
n-Propyl alcohol	19.5	3.45	0.28	6.42	5.25
n-Butyl alcohol	—	—	—	—	5.08
Propylene glycol	—	—	16.0	16.0	—
Ethylene glycol	0.81	1.10	1.68	7.00	—
Ethyl ether	16.6	5.75	20.8	17.8	—
Dioxane	0.79	3.00	0.42	20.00	4.18
Aniline	0.51	0.59	0.44	15.2	—
Dimethyl formamide	0.05	4.30	0.62	7.75	—
Benzene	15.1	7.16	12.1	11.0	7.75
Toluene	—	—	—	—	4.96
Xylene	—	—	—	—	6.67
Cyclohexane	18.3	5.82	2.66	10.2	3.57
Acetic acid	0.98	2.92	1.68	0.00	—

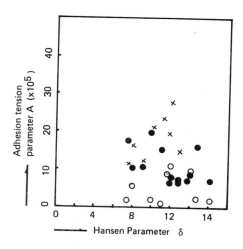

FIGURE 8.5 Relations between adhesion parameter A and Hansen parameter δ. Iron oxide (o), phthalocyanine blue (•), insoluble azo. B (x).

TABLE 8.7
Adhesion Tension Parameters Estimated by
Equation (7)
($\times 10^3$ g/cm^2)

	Iron oxide	Titanium dioxide	Molybdate orange	Insoluble azo. A	Insoluble azo. B
Methyl alcohol	7.97	0.29~0.42	3.03	5.82	3.55
Ethyl alcohol	6.02	0.06	9.30	9.30	3.31
n-Propyl alcohol	7.31	0.55	4.63	4.63	4.36
n-Butyl alcohol	—	0.46	—	—	1.88
Propylene glycol	1.75	—	—	—	7.85
Ethylene glycol	1.38	—	3.86	3.86	2.53
Ethyl ether	1.66	—	10.2	10.17	4.97
Dioxane	1.35	9.34	5.45	5.45	5.23
Aniline	0.29	—	4.31	4.31	6.15
Dimethyl formamide	9.08	—	2.41	2.41	2.75
Benzene	9.08	9.90	9.56	9.56	—
Toluene	—	3.11	—	—	—
Xylene	—	11.8	—	—	—
Cyclohexane	4.38	7.8	4.77	4.70	3.60
Acetic acid	7.40	—	3.07	3.07	4.90

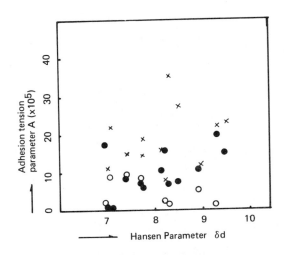

FIGURE 8.6 Relation between adhesion parameter A and Hansen parameter δ_d. Iron oxide (o), phthalocyanine blue (•), insoluble azo. B (x).

TABLE 8.8
Adhesion Tension Parameters Estimated by Equation (7)
($\times 10^3$ g/cm^2)

	Isoindoli- none	Quinacri- done	Dioxazine	Phthalo- cyanine blue	Carbon black
Methyl alcohol	8.13	5.27	3.28	5.73	13.3
Ethyl alcohol	16.5	1.84	0.29	4.64	18.7
n-Propyl alcohol	23.5	2.08	0.10	4.17	16.4
n-Butyl alcohol	—	—	—	—	15.9
Propylene glycol	—	—	—	10.4	—
Ethylene glycol	0.98	0.66	0.58	4.55	—
Ethyl ether	16.9	3.46	7.12	11.6	—
Dioxane	0.95	1.80	0.14	13.0	13.1
Aniline	0.61	0.36	0.15	9.87	—
Dimethyl formamide	0.06	2.59	0.21	5.03	—
Benzene	18.2	4.31	4.14	7.14	24.2
Toluene	—	—	—	—	15.5
Xylene	—	—	—	—	20.8
Cyclohexane	22.0	3.51	0.91	6.62	11.2
Acetic acid	1.18	1.76	0.57	—	—

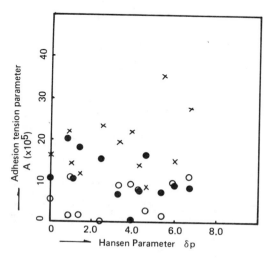

FIGURE 8.7 Relation between adhesion parameter A and Hansen parameter δ_p. Iron oxide (o), phthalocyanine blue ($\bullet$), insoluble
azo. B (x).

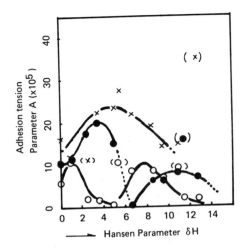

FIGURE 8.8   Relation between adhesion parameter A and Hansen parameter δ_H. Iron oxide (o), phthalocyanine blue (●), insoluble
azo. B (x).

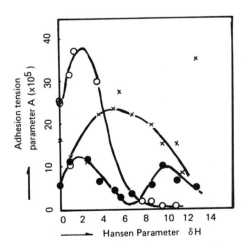

FIGURE 8.9   Relation between adhesion parameter A and Hansen parameter δ_H. Type A: mono peak (Titanium dioxide (o)), Type
B: broad peak (insoluble azo. B (x)), Type C: two peak
(insoluble azo. A (●)).

(1) Type A which has a comparatively explicit maximum peak;
(2) Type B which gives a rather broad peak; and
(3) Type C which has two distinct peaks.

Correlation Between Dispersability and
Adhesion Tension Parameter

As discussed before, though it is not possible to compare adhesion tensions of pigments precisely in our experiments, but the porosity mean capillary radius is not so different from hydrodynamic mean radius so that the results obtained may be qualitatively comparable. This consideration is supported by our results as summarized in Tables 8.7 and 8.8. In accordance with this viewpoint, the spectrum characteristics and dispersabilities of various pigments in two types of vehicles are summarized in Table 8.9.

Pigments were dispersed in two vehicles under the same conditions, namely a xylene solution of butyrated melamine resin and a xylene solution of maleic resin. Dispersabilities are expressed as the ratio of the Stokes mean radius of particles to the mean radius of the primary particles of the pigment used. The smaller value means better dispersability.

As shown in Table 8.10, the pigments (samples No. 5 and 8) having low levels of maximum adhesion tension parameter in the spectrum were hard to disperse in both vehicles. Pigments which had Type A spectrum were easily dispersed in the butyrated melamine vehicle but were difficult in the maleic resin vehicle. Pigments which had Type C spectrum were easily dispersed in both vehicles except for carbon black and phthalocyanine blue. The carbon black and the phthalocyanine blue pigment had Type C spectrum, but both pigments were similar to Type A pigments in dispersion behavior. The reason for this irregularity cannot be readily explained from these experiments. The results mentioned above suggest that the interaction between polymers and pigments is similar to that of between solvents and pigments.

Discussion

Since the interfacial tension ($\gamma_{s\ell}$) between pigments and solvents is positive, the adhesion tension of pigment with solvent must be maximum when $\gamma_{s\ell}$ is zero according to equation (12)

$$W_A = \gamma_\ell \cos \Theta = \gamma_s - \gamma_{s\ell} \tag{12}$$

where W_A is the adhesion tension, γ_ℓ and γ_s are the surface tensions of solvent and pigment, respectively, and $\gamma_{s\ell}$ is the pigment/solvent interfacial tension.

TABLE 8.9
Classification of Pigments by the Spectrum

Type	Character	Pigments	The position of peaks (δH) First	Second
A	Mono peak	Titanium dioxide	3.0	—
B	Broad peak	Insoluble azo. B	5.5	
C	Two peaks	Molybdate orange	2.0	8.5
		Iron oxide	1.0	8.0
		Insoluble azo. A	2.0	9.2
		Quinacridone	1.0	11.0
		Dioxazine	2.0	11.0
		Isoindolinone	0.0 >	9.5
		Carbon black	1.0	9.0
		Phthalocyanine blue	3.5	11.0

Experimental and theoretical investigations of $\gamma_{s\ell}$ have been made by Sell-Neumann [7], Fowkes [8], Wu [9], and Kitazaki and Hata [10], and the relationship between interfacial tension ($\gamma_{s\ell}$) and surface tension of pigment (γ_s) and solvent (γ_ℓ) has been expressed by equation (13).

$$\gamma_{s\ell} = (\sqrt{\gamma_s^a} - \sqrt{\gamma_\ell^a})^2 + (\sqrt{\gamma_s^b} - \sqrt{\gamma_\ell^b})^2 + (\sqrt{\gamma_s^c} - \sqrt{\gamma_\ell^c})^2 \qquad (13)$$

$$\gamma_s = \gamma_s^a + \gamma_s^b + \gamma_s^c \qquad \gamma_\ell = \gamma_\ell^a + \gamma_\ell^b + \gamma_\ell^c$$

γ^a: Dispersive parts of the surface tension
γ^b: Polar parts of the surface tension
γ^c: Hydrogen bonding parts of the surface tension

As evidenced by equation (13), the interfacial tension become minimal when the surface tensions of pigments and solvents are identical, and consequently the adhesion tension maximizes. From this we may conclude that our results obtained imply that the interaction between pigments and solvents is dominated by hydrogen bonding.

Accordingly, we suppose that the hydrogen-bonding property of a pigment is identical with that of the solvent which gives the maximum

TABLE 8.10

Correlations Between Adhesion Tension Parameter
and Dispersability in Vehicles

Sample No.	Pigments	Adhesion tension parameter at peaks ($\times 10^3$ g)		Dispersability in vehicles	
		First peak	Second peak	dm/dp[a] Melamine	dm/dp[a] Maleic resin
1	Iron oxide	9.0	7.2	6.2	1.2
2	Titanium dioxide	11.8		4.3	1.0
3	Molybdate orange	10.2	4.6	1.2	1.2
4	Insoluble azo. A	10.4	6.6	1.0	2.1
5	Insoluble azo. B	5.2		8.0	7.0
6	Isoindolinone	22.0	16.5	—	—
7	Quinacridone	18.2	3.6	1.6	1.6
8	Dioxazine	7.5	3.5	7.0	7.5
9	Phthalocyanine blue	13.2	5.8	5.2	1.0
10	Carbon black	19.8	18.8	1.0	8.0

[a] dm: Mean radius in vehicles determined by disc centrifuge method (Joyce Loeble MK-3).
dp: Mean radius of primary particles shown in Table 8.1.

adhesion tension in the spectrum shown in Figure 8.8 and Figure 8.9. It can easily be understood that the adhesion tension spectrum changes with pigments as shown in Table 8.9. It seems reasonable to consider that a low value of adhesion tension means a low density of hydrogen-bonding sites at the pigment surface.

Thus, it may be said that the concept of "adhesion tension spectrum" can be established with theoretical as well as experimental backing; Table 8.10 demonstrates clearly the existence of distinctive correlation between the type of the spectrum of pigment and its dispersability in vehicles.

Sorensen [6] proposed that all pigments, resins, and solvents can be classified according to the acid-base concept and an acidic pigment (electron acceptor) is easily dispersed in a basic vehicle (electron donor) but is hardly dispersible in an acidic vehicle, and vice versa.

Referring to Sorensen's classification, butylated melamine resin is an electron donor (basic), while maleic resin is an electron acceptor (acidic). Therefore, as is evident from Table 8.10, the titanium dioxide pigment used in this experiment should be basic, while carbon black which exhibits an opposite tendency must be acidic according to Sorensen's definition.

In a A-δH diagram, titanium dioxide shows a Type A spectrum and has a maximum adhesion tension with a solvent which has a relatively small hydrogen-bonding parameter (δH). It is concluded that this type of pigment may have electron donating moieties on its surface, while a pigment which also has a Type A spectrum but a maximum at large δH would have electron accepting groups and behaves as an acidic pigment. Many other pigments examined here have Type C (two peaks) spectrum as shown in Table 8.10. It is natural to assume that they have more or less amphoteric surface characteristics.

Another aspect worthy of attention is the maximum value of adhesion tension. When a pigment exhibits a small adhesion tension maximum value, it is considered to have a low hydrogen-bonding tendency and thus be an inactive pigment. Insoluble azo B and dioxazine in Table 8.10 are thought to fall in this category and disperse rather poorly in both vehicles.

Our experimental results and the reasoning described above are in agreement with the work of Sorensen and our earlier report [11], except for the carbon black and the phthalocyanine blue.

APPLICATIONS

Color Stability of Paint and Dispersion Behavior
of the Pigments Employed

The color change of a paint containing a combination of different pigments during storage is an often-encountered difficulty for paint formulators. The relation between the dispersion behavior of component pigments and the color stability of enamels based on maleic resin solution was investigated.

As is well-known, the cuase of color instability [12] in mixed pigmented paints is coflocculation (heterocoagulation) between pigments. This defect can be easily evaluated by visual inspection of paints in glass cylinders. The results are compiled in Table 8.11. In the table, pigments are divided into groups according to their behavior concerning the A-δH spectrum. The degree of heterocoagulation of each combination of pigments are expressed in the corresponding lattice. It is clear that coflocculation occurs only sparingly with the combination of pigments which come from same group, while the dispersion stability is generally poor when the component pigments belong to different groups.

TABLE 8.11

Dispersion Stabilities Against Heterocoagulations of Maleic Resin Enamels in Storage and the Wettability Properties of Pigments

| | | Type | | | | | | | | | |
| | | C | | | | | A-2 | | A-1 | B | |
Type	Enamel	Iron oxide	Molybdate orange	Quinacri-done	Isoindoli-none	Insoluble azo. A	Titanium dioxide	Phthalo-cyanine blue	Carbon black	Insoluble azo. B	Dioxazine
C	Iron oxide	o	o	o	o	x	o	x		x	x
	Molybdate orange	o	o	o	o	x	o	x		x	o
	Quinacridone	o	o	Δ	Δ	o	o	o		x	Δ
	Isoindolinone	x	x	o	x	x	x	x		x	x
	Insoluble azo. A	o	o	o	x		x	x		o	x
A-2	Titanium dioxide	x	o	o	x	x	o	o		x	o
	Phthalocya-nine blue		x	o	x	x				x	o
A-1	Carbon black										
B	Insoluble azo. B	x	x	x	x	o	x	x			x
	Dioxazine	x	o	Δ	x	x	o	o		x	x

*1. C: Type C in the spectrum, Amphoteric; A-2: Type A in the spectrum, Basic, A-1: (Type A in the spectrum, Acidic); B: Type B in the spectrum, Nonactive.

*2. Dispersion stability against heterocoagulation was estimated from the storage stability for 4 weeks at 40°C. O: good, Δ: fairly good, X: poor.

TABLE 8.12

Correlation Between the Dispersion Stability of
Poly-Vinyl Chloride Powders[a] in Solvents and
Acid-Base Properties of Solvents

	Properties of solvent		Properties of dispersion[b]		
	Chemical formula	Acidity	Viscosity (C. P.)	Dispersion stability	Swelling ratio
Ethylene glycol mono-ether	ROROH	Amphoteric	10-15	Very good	1.2~1.6
Alcohol	ROH	Acid	150-300	Poor	1.0~1.15
Aromatic hydro-carbon	⟨⟩-R	Base	200-600	Poor	1.7~2.0
Ketone	RCOR	Base	8000<	Good	5.0~2.0
Ester	RCOOR	Base	700<	Good	2.5~10

[a] Poly-vinyl chloride powder: Mitubishi Monsanto Co., Ltd., p. 410.

[b] Volume fraction of P.V.C. powders in dispersions: 0.315.

Again, our concept of extended acid-base interaction, i.e., adhesion tension spectrum is useful in controlling the coflocculation of pigments and paint color stability in storage.

Dispersion Stability of PVC Powder in Solvents

It is known that a PVC resin dispersion (organosol and plastisol) exhibits a variety of stability depending on the dispersing media. To make clear this phenomenon, the dispersion stability of PVC powder was examined in a series of solvents. The results obtained are summarized in Table 8.12, where it is obvious that the swelling ratio and the dispersion stability is dominated by the acid-base characteristics of solvents. As PVC resin is acidic, it swelled remarkably in basic solvents, while acidic solvents such as alcohols are poor swelling agents for the resin. It is quite noteworthy that monoether of ethylene glycol, which is amphoteric in character, is an excellent disperser for PVC resin in spite of its low swelling ability. This is interpreted as a result of the combination of two factors, that is, adsorption of the basic group and a steric stabilization effect by the relatively large molecule of the solvent.

REFERENCES

1.   Tanaka, K. and Koishi, M. J. Jpn. Soc. Color Mater. 49:22, 1976.
2.   Tanaka, K. and Koishi, M. J. Jpn. Soc. Color Mater. 49:473, 1976.
3.   Tanaka, K. and Koishi, M. J. Jpn. Soc. Color Mater. 50:37, 1977.
4.   Kubo, T., Suitou, E., Nakagawa, Y. and Hayakawa, S. "Funtai (Powder)," Maruzen, p. 226, 1961.
5.   Hansen, M. and Skaarup, K. J. Paint Technol. 39(511):511, 1967.
6.   Sorensen, P. J. Paint Technol. 47(602):31, 1975.
7.   Sell, P. J. and Neumann, A. W. Angew. Chem. 78:321, 1966.
8.   Fowkes, F. M. J. Phys. Chem. 66:382, 1968.
9.   Wu, S. J. Adhesion 5:39, 1973.
10. Kitazaki, Y. and Hata, T. J. Adhesion Soc. Jpn. 8:131, 1972.
11. Fujitani, T. and Satoh, T. Shikizai Kyokai Congress Book, p. 29, 1980.
12. Nakatogawa, Y. and Satoh, T. J. Jpn. Soc. Color Mater. 49:725, 1976.
13. National Paint and Coating Association, "Raw Material Index Pigment Section," 1975.

THE ROLE OF POLYMERS IN CONTROLLING
DISPERSION STABILITY

Brian Vincent

Department of Physical Chemistry
University of Bristol
Bristol, United Kingdom

INTRODUCTION

It is now firmly established that polymers play a major role in controlling the stability behavior of colloidal dispersions [1, 2]. This fact must have been known, albeit inadvertently, to some of the ancient civilizations who made, for example, crude pigment dispersions stabilized by natural gums and proteinaceous matter. Faraday's famous gold sol, preserved at the Royal Institution in London, is stabilized by added "organic matter." This protective role against aggregation imparted by macromolecules arises from the adsorption of these molecules at the particle/solution interface, and is now referred to as "steric stabilization." It is made use of in a wide variety of industrial formulations where stable dispersions or emulsions are required. It is particularly useful in the case of nonaqueous dispersions, or for aqueous dispersions where a high bulk electrolyte concentration is present because under these circumstances charge stabilization is normally inadequate.

There are many examples of industrial products and processes, on the other hand, where flocculation of an otherwise stable dispersion is required (e.g., water purification, mineral processing, harvesting of bacteria in biotechnological processes, etc.). In other cases, gross flocculation is not required, but rather a controlled degree of weak, reversible flocculation. Examples here would include the formulation of dispersions possessing thixotropic and/or anticaking properties. Pigment dispersions (e. g., in paints or printing inks) frequently fall into this category. This paper will primarily be concerned with reviewing the ways in which weak flocculation of this type can be achieved. It will begin by reviewing the theoretical background to the subject, and then go on to illustrate this with some experiments on model systems. Many of the examples will be based on experimental work carried out in recent years at Bristol University.

THEORETICAL BACKGROUND TO WEAK FLOCCULATION

In discussing the stability of dispersions towards aggregation, a careful distinction has to be made between stability in the thermodynamic and kinetic senses. A dispersion is thermodynamically stable if aggregation would lead to an increase in the free energy of the system. If a lower free energy would result then the system is either unstable or metastable; it is metastable if a free energy barrier determines the rate of aggregation; if it is sufficiently high, then the dispersion may be "stable" (in the kinetic sense) for a long period. This situation is the one which pertains for charge-stabilized dispersions at low electrolyte concentrations. The theoretical analysis of this type of stability has its foundations in the so-called D. V. L. O. theory [3].

The typical form of the pairwise interparticle-interaction free energy curve $G_i(h)$, is shown in Figure 9.1(a) for a charge-stabilized dispersion. The shape arises from the "soft" i. e., long-range, nature of the repulsive, electrical-double-layer term (G_E). For a sterically-stabilized dispersion the basic form of the $G_i(h)$ curve is significantly different. As can be seen Figure 9.1b), it is characterized by a shallow minimum (G_{min}), rather than the maximum (G_{max}) that occurs in the charge-stabilized case. This minimum results from the "hard" i. e., short-range, nature of the steric interactions. Clearly, the form of $G_i(h)$ is more complex if both charge and steric-stabilization are present in a given dispersion. However, we shall not be concerned with systems of this type here, although they do occur quite frequently in practice. In most of the experimental work to be described later, charge stabilization is either absent, or effectively eliminated by the presence of high concentrations of electrolyte.

The origin of the steric interaction (G_s) is complex [1, 2] and will not be reviewed in detail here. A number of theoretical approaches [2] have been developed for predicting the form of G_s. Some insight may be gained by examining an expression derived by Smitham et al. [4] for the steric interaction between two spherical particles (radius $\underline{a}$), covered by an adsorbed layer of polymer molecules (thickness δ):

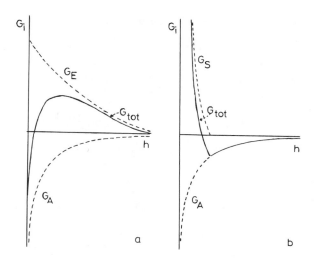

FIGURE 9.1   Particle interaction free-energy diagrams for: (a) charge-stabilized dispersions; and (b) sterically-stabilized dispersions. G_A is the van der Waals term, G_E the electrical double-layer term, and G_S the steric term. G_{tot} is the total interaction. h is the shortest distance between particle surfaces.

$$G_s = \frac{2 \pi a \; kT \; V_s^2 \; (\Gamma_2 n)^2}{V_1} \; (\tfrac{1}{2} - \chi) \cdot S_{mix} + 2 \pi a \; kT \; \Gamma_2 \cdot S_{el} \qquad (1)$$

$$\underbrace{\hphantom{xxxxxxxxxxxxxxxxxxxxxxxxxxxxxxxxxx}}_{(G_{mix})} \qquad\qquad \underbrace{\hphantom{xxxxxxxxxxxxxxx}}_{(G_{el})}$$

where Γ_2 is the number of chains per unit area of surface; n the number of segments per chain; V_s the volume of a segment; V_1 the volume of a solvent molecule; χ the Flory polymer-solvent interaction parameter; k the Boltzmann constant; and T the absolute temperature. S_{mix} and S_{el} are geometrical functions i. e., functions of a , δ and h, and also the form of the segment density distribution in the polymer sheath, and refer to the "mixing" term and "elastic" term, respectively. The mixing term has to do with the local buildup of segment concentration in the interaction zone between the two overlapping or interpenetrating polymer layers (h < 2δ). For the segments in a good (better-than-theta) solvent environment ($\chi < \tfrac{1}{2}$), this corresponds to a local increase in osmotic pressure which opposes the overlap of the polymer layers, i. e., a repulsive term. However, in a bad (worse-than-theta) solvent ($\chi > \tfrac{1}{2}$), the leading term on the r. h. s. of equation (1) is negative, implying an attractive term. The elastic term remains repulsive under all conditions and has to do with the loss in configurational entropy of the whole chains, as a result of their compression in the interaction zone. For this reason this term is also sometimes referred to as the "volume-restriction term." One of the difficulties in discussing the elastic term is whether it is effective for $h < 2\delta$ or only for $h < \delta$ [4, 5]; this decides the form of S_{el}. For high molecular weight, terminally-anchored chains (tails) it may be reasonable to assume that significant compression of the chains only occurs when the chains attached to one surface come into contact with the actual surface of the opposing particles, i.e., the elastic term is only important for $h < \delta$. With low molecular weight tails or absorbed homopolymer (loop and train conformation), this assumption becomes invalid [2, 6] and interaction occurs at $h < 2\delta$. One other feature to note from equation (1) is that G_{mix} depends on $(\Gamma_2)^2$ whereas G_{el} depends only on Γ_2.

For sterically-stabilized dispersions the other main parameter, in addition to G_{min}, which controls their stability/instability behavior is the particle number density (ρ) or volume fraction (Φ). (N. B. $\Phi = \rho v$, where v is the particle volume). In this respect, sterically-stabilized dispersions resemble molecular systems. Thus, just as one may express the pressure, p, of a vapor as a virial expansion of the form

$$p = \rho kT + B\rho^2 + \dots \dots \tag{2}$$

(where ρ is now the number of molecules per unit volume and B the second virial coefficient), so one may express the osmotic pressure, π, of a dispersion as

$$\pi = \rho kT + B\rho^2 + \dots \dots \tag{3}$$

where the first term on the r.h.s. is the ideal term and the second term is the leading nonideal term. B is, in this case, a function of the pairwise interactions between the particles, i.e., G_{min}. At high ρ, the terms in ρ^3 etc. become increasingly more significant and multibody interactions become important.

Just as with a vapor, there exists (at a given temperature) a critical saturation pressure, below which the vapor is thermodynamically stable, so by analogy, there should exist for a weakly-interacting particulate dispersion a critical particle number concentration (or volume fraction), below which the dispersion is stable to flocculation. Above this critical flocculation volume fraction (c. f. Φ) the dispersion should flocculate reversibly. This begs the question: what is the nature of the equilibrium state that is achieved? Does one see a distribution of singlets, doublets, triplets, . . . multiplets, or does one see, primarily, singlet particles in equilibrium with large multiplets i.e., flocs? The experimental evidence, as we shall see, favors the latter; this again is in accord with what we observe with vapor/liquid equilibria, where a critical nucleus size exists (for a given supersaturation pressure and temperature), only above which do the nuclei grow into large liquid droplets which coalesce and separate out (under gravity) as a separate liquid phase.

The implication of the above discussion is that in order to achieve control of the extent (and indeed the strength) of flocculation in a sterically-stabilized dispersion one must control G_{min} and Φ. In practice, the range of particle volume fraction is normally fixed by the usage to which the dispersion is put. Therefore one must be able to control G_{min}. Before discussing this it is important to emphasize that the polymer must be present at the surface at high coverage and be strongly anchored to the surface. If bare patches exist, on either surface, then polymer bridging may result [1, 2, 7], which generally results in strong, open-textured flocs being formed. Alternatively, one may actually find two particle surfaces coming into direct contact, resulting in irreversible coagulation. This would also result if the polymer molecules are so weakly adsorbed that desorption may occur during particle adhesion. As was pointed out earlier this may be desirable in certain applications, but is not the requirement in many others, where, as we have seen, controlled, weak (reversible) flocculation is desired.

Long et al. [8] and more recently, Vincent et al. [9] have discussed the analogy between weak flocculation and phase separation. They have suggested that the equilibrium state of flocculation may be represented by a Boltzmann relationship of the form

$$\Phi^d = \Phi^f \exp \left[\frac{zG_{min}}{2kT} \right] \tag{4}$$

where Φ^d is the volume fraction of particles in the disperse (i. e. , singlet) phase; Φ^f the volume fraction of particles in the floc phase; z is the co-ordination number of a particle in a floc; therefore, both z and Φ^f depend on the floc morphology (for random close-packing z ~ 8 and Φ^f = 0. 66; for hexagonal close-packing, z=12 and Φ^f=0. 74). Again, by analogy with molecular phase changes, we may identify c.f. Φ with Φ^d. Thus, equation (4) gives us the desired relationship between Φ and G_{min}. This is shown in Figure 9. 2 for two types of floc structure.

There are several ways in which G_{min} may be controlled in practice: (i) by control of δ, the thickness of the polymer layer; (ii) by control of χ, i. e. , the solvency of the continuous phase for the adsorbed polymer molecules; and (iii) by control of the concentration of free polymer in solution. In some cases a combination of these effects may occur.

Control of δ is normally achieved by the choice of molecular weight, for a given adsorbed polymer type. One is making use here of the fact that the van der Waals interaction G_A is longer-range than G_S (Figure 1b). In a good solvent environment ($\chi < \frac{1}{2}$) and at high coverage (Γ_2) the mixing

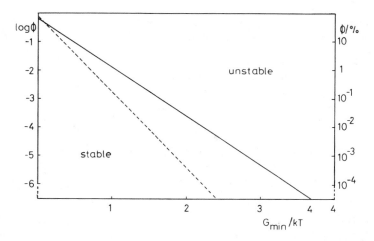

FIGURE 9.2   Theoretical particle volume fraction (Φ) – interaction free
energy minimum (G_{min}) diagram for weakly interacting
particles, based on equation (4) [8,9] —for randomly close-
packed flocs; - - - for hexagonally close-packed flocs.

term (equation 1) is strongly repulsive, as is the elastic term. G_S therefore rises very steeply at $h < 2\delta$ and, to a first approximation, under
these conditions we can regard G_{min} as being the value of G_A at $h = 2\delta$.
One may then estimate G_{min} for one of the equations for G_A [1,3]. The
simplest of these is the approximate form given originally by Hamaker [10]:

$$G_A = -\frac{(A_1^{\frac{1}{2}} - A_2^{\frac{1}{2}})^2 a}{12h} \tag{5}$$

where A_1 and A_2 are the Hamaker constants for the particle and continuous
phase, respectively. Use of equation (4) implies that the Hamaker constant
of the adsorbed polymer sheath (A_3) is the same as that of the continuous
phase i.e., $A_3 = A_1$. This is a reasonable assumption in the case of high
molecular weight adsorbed polymers, where even at high coverages, a
major portion of the volume of the sheath is occupied by solvent molecules.
Where A_3 differs significantly from A_1, one of the more complex expressions for G_A is required, e.g., those derived by Vold [11] and Vincent
[12].

By way of illustration of the use of equation (4) consider the following
examples. In order to achieve $|G_{min}| = 1$ kT (corresponding to a c.f. Φ
of ~ 0.012 (Figure 9.2), for 100 nm radius polystyrene particles in water,

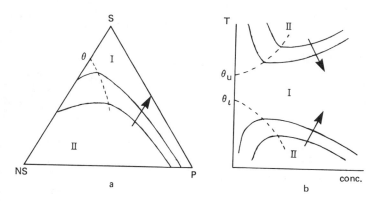

FIGURE 9.3   Schematic phase diagrams for nonaqueous polymer solutions:
(a) polymer (P) + solvent (S) + nonsolvent (NS), at constant
temperature and pressure; (b) temperature-composition
(polymer + solvent) phase diagram at constant pressure.
--- locus of consolute (critical) points, which intersects the
S-NS or T axis at θ-conditions. Arrows indicate direction
of increasing molecular weight.

δ has to be 7.6 nm and 18.8 nm for 100 nm radius silica particles in
toluene.[*]
    The second way of controlling G_{min} is through variation in χ, the
polymer-solvent interaction parameter. χ is in fact a complex function of
the temperature [13], as well as the nature of the polymer and the solvent.
As we have already seen, however, if $\chi > \frac{1}{2}$ then the mixing contribution
to the steric interaction (G_{mix}) becomes attractive. This condition may be
conveniently achieved by adding nonsolvent to the system (including electro-
lytes in the case of aqueous dispersions), and/or by a suitable adjustment
of the temperature. A useful guide here is the form of the polymer +
solvent + nonsolvent (constant temperature) phase diagram, or the
temperature-composition phase diagram, as appropriate. These are
illustrated, schematically, in Figure 3(a) and (b), respectively. Note that
these are typical for nonaqueous systems; even so in many cases the upper-
θ temperature (θ_u, Figure 3b) is not observed, either because it occurs
above the normal boiling point of the solvent or because the polymer
degrades. For most aqueous systems, on the other hand, only an upper-θ
temperature is observed.

----

[*] The Hamaker constants used in these calculations were based on the values
given by Vincent [12].

In the case of solvent + nonsolvent mixtures, $\chi > \frac{1}{2}$ for nonsolvent volume fractions greater than the Θ-composition, while for variable temperatures, $\chi > \frac{1}{2}$ for $\Theta_\ell > T > \Theta_u$.*

Although G_{mix} will be negative when $\chi > \frac{1}{2}$, the value of G_{min} depends on the net effect of G_{mix}, G_{el} and G_A. As discussed previously, G_{mix} only dominates the total interaction for terminally-anchored, high molecular weight chains (where G_{el} and G_A are negligible for small degrees of interpenetration). The situation is illustrated schematically in Figure 8.4. In the case of high molecular weight tails (Figure 9.4a) G_{min} increases dramatically in the region of $\chi = \frac{1}{2}$; one thus expects to see a strong correlation between critical flocculation conditions and Θ-conditions. For short tails (or loop and train configurations, e.g., adsorbed homopolymer or random copolymers) (Figure 9.4b), G_{el} offsets the effects arising from the change of sign of G_{mix} much more effectively. Also effects associated with G_A may play a significant role, i.e., a decrease in solvency leads, in general, to a decrease in δ (and also to an increase in the Hamaker constant of the adsorbed layer), thus increasing $|G_{min}|$ through increase in G_A. Clearly, in these cases, the effect of change in χ on G_{min} are much more difficult to predict. Certainly, a strong correlation between critical flocculation conditions and Θ-conditions can no longer be expected. These points will be illustrated in the experimental section.

The third way in which G_{min} may be controlled is through the addition of free polymer to the continuous phase. The manner by which this occurs is rather subtle and not, as yet, completely understood. Several theories have recently [9, 15-17] been proposed which purport to take account of the effect of the presence of free polymer in solution on interparticle interactions. The theories of Vrij [15], Joanny et al. [16] and Feigin and Napper [17] are all based on the concept that at polymer concentrations beyond which overlap of the polymer coils in solution occurs, those coils in contact with (although not attracted by) the surface of a colloidal particle have a lower configurational entropy and therefore, a higher free energy than those in bulk. Hence, because particle aggregation leads to a net displacement of such coils (into bulk solution) from the neighborhood of the interaction zone between two approaching particles, there is a lowering in the free energy of the system if the particles come into contact. The theory of Vincent et al. [9], on the other hand, takes a somewhat different approach. Here, the net free energy balance is calculated based on the changes in segment-solvent mixing free energy. It is shown that for small degrees of overlap of the two (polymer plus solvent) sheaths surrounding two approaching colloidal particles, there can be a net decrease in the free energy of the system as a result of the displacement

---

*
For simplicity we neglect the polymer concentration dependence of χ; Evans and Napper [14] have considered the inclusion of this effect in the theory of steric interactions.

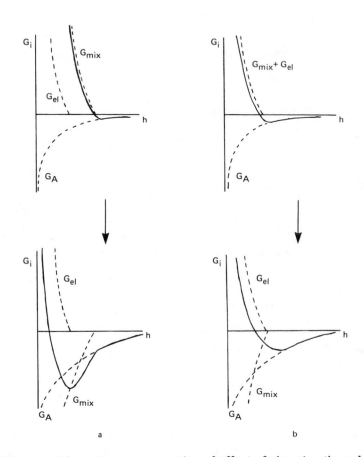

FIGURE 9.4   Schematic representation of effect of changing the solvency
($\chi < 0.5 \rightarrow \chi > 0.5$) for sterically-stabilized dispersions:
(a) for high-molecular weight terminally-anchored chains;
(b) for low-molecular weight terminally-anchored chains
(or adsorbed homopolymer). G_{mix} is the mixing term,
G_{el} the elastic term, G_A the van der Waals term; —— the
total interaction.

of those coils (into bulk solution) which were previously overlapping the sheaths in the interaction zone. In both cases the decrease in free energy constitutes an attractive force between the particles, leading to an increase in $|G_{min}|$. Again, examples of this effect will be given in the next section.

EXPERIMENTAL ASPECTS OF WEAK FLOCCULATION

In this section experiments are described which illustrate some of the principles discussed in the previous section, in particular the effects of variation of particle volume fraction (Φ), the thickness of the adsorbed polymer sheat (δ), the Flory interaction parameter (χ), and the concentration of free polymer, on the stability/flocculation behavior of sterically-stabilized dispersions.

The first set of experiments that demonstrated the existence of a critical flocculation volume fraction (c.f. Φ) for sterically-stabilized dispersions were those of Long et al. [8]. These authors studied the system containing polystyrene latex particles plus adsorbed dodecylhexaethoxylate ($C_{12}E_6$) at monolayer coverage in aqueous solution. Charge interactions were removed by the addition of 0.1 mol dm^{-3} $BaCl_2$. The results are shown in Figure 9.5 in the form of n versus Φ plots: n is a light scattering parameter (n = dlog τ/dlog λ, τ = turbidity, λ = wavelength) which is an effective measure of average particle size; a decrease in n corresponds to an increase in particle size. Thus, the break points in Figure 9.5 correspond to the onset of flocculation in the dispersions. The different curves refer to different lattices having the particle radius indicated. It can be seen that the c.f. Φ decreases as the particle size, and therefore $|G_{min}|$, increases. This is in accord with equation (4).

In the above experiments $|G_{min}|$ was varied by varying the particle radius, keeping δ fixed. Some experiments which illustrate the effect of varying δ, keeping the radius constant, are those of Lambe [18]. In this work different molecular weight fractions (1500 to 20,000) of polyethylene oxide PEO were adsorbed onto polystryrene latex particles at monolayer coverage in aqueous solution, and the electrolyte concentration then adjusted to 10^{-2} mol dm^{-3} $Ba(NO_3)_2$, again to eliminate electrical double layer interactions. The results are shown in Figure 9.6 in the form of n versus log Φ plots. The c.f. Φ decreases now with decreasing PEO molecular weight, i.e., decreasing δ, again corresponding to an increase in $|G_{min}|$. Note, that no break point was observed with PEO 20,000 over the range of Φ values that could be studied using the turbidimetric technique, although one should exist at a higher Φ value.

The effect of varying χ is illustrated by the results presented in Figure 9.7. Here the critical flocculation temperature (c.f. T) for two polystyrene (PS) latices carrying terminally-anchored PEO chains (tails) of molecular weight 750 and 2000, respectively, are shown as a function of log Φ, for different $MgSO_4$ concentrations [6]. The c.f. T. values

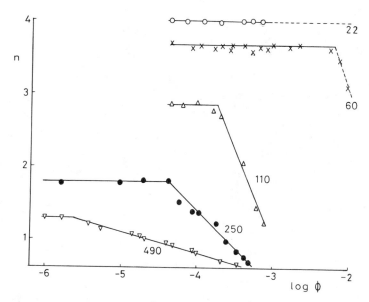

FIGURE 9.5   n (= dlog τ/dlog λ) versus log Φ plots for PS latices plus
adsorbed $C_{12}E_6$ plus 10^{-1} mol dm^{-3} $Ba(NO_3)_2$ [8]. Numbers
by the lines indicate particle radius of singlets (nm) . 25°C.

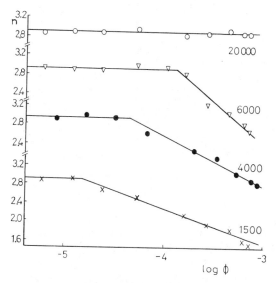

FIGURE 9.6   n (= dlog τ/dlog λ) versus log Φ for PS latex plus adsorbed
PEO plus 10^{-2} mol dm^{-3} $Ba(NO_3)_2$ [18]. Numbers by the
lines indicate PEO molecular weight . 25°C.

179

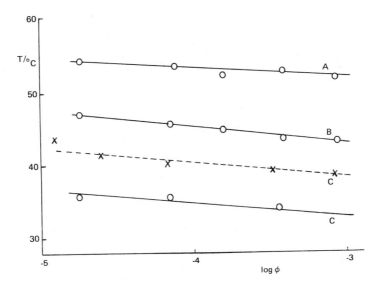

FIGURE 9.7   c.f.T. versus log Φ plots for PS–PEO 750 (——) latex and
PS–PEO 2000 (---) latex, at different MgSO$_4$ concentrations: (a) 0.065 mol dm^{-3} ($\ominus$ = 85°C); (b) 0.163 mol dm^{-3}
($\ominus$ = 72°C); (c) 0.26 mol dm^{-3} ($\ominus$ = 59°C) [6].

were again obtained from log τ/log λ plots [6]. With the PS–PEO 750
latex it can be seen that the c.f.T. (at a given Φ) decreases as the MgSO$_4$
concentration increases (i.e., χ increases), implying $|G_{min}|$ also
increases at any fixed temperature with increasing electrolyte concentration. It is of interest to compare the c.f.T. values with the corresponding
$\ominus$-temperatures (see legend to Figure 9.7) for a given electrolyte concentration. Consider, for example, 0.26 mol dm^{-3} MgSO$_4$; the $\ominus$-temperature
for PEO in this solution is 59°C [19]. The c.f.T. for PS–PEO 750 latex is
36 ±2°C and that for PS–PEO 2000 is 41 ±2°C, clearly well below the
$\ominus$-temperature. On the other hand, Napper [20] has shown that when the
molecular weight of the PEO tails is somewhat higher (> ~5000) there is a
close correspondence between the c.f.T. and the $\ominus$-temperature. This
exemplifies the point made in the theoretical section that with high molecular weight tails, the G_{mix} term dominates the interparticle interactions,
and, hence, the change in sign of G_{mix} leads to flocculation under worse-
than-$\ominus$ conditions (c.f. Figure 9.4a). As the molecular weight of the PEO
tails (and hence δ) decreases, so effects from G_{el} and G_A begin to play a
greater role as discussed (c.f. Figure 9.4b), hence the increasing lack of
correlation between the c.f.T. and the $\ominus$-temperature.

The other feature apparent from Figure 9.7 is the small, but real, Φ dependence of the c.f.T. This feature has been investigated in more detail recently by Cowell and Vincent [21] who extended the range of Φ values to 0.1. The c.f.T. was determined at these much higher Φ values using a zero-shear viscometry technique. The results for PS-PEO 2000 in 0.26 mol dm^{-3} MgSO$_4$ are shown in Figure 9.8. Here it can be seen that there is a definite dependence of the c.f.T. on Φ. Indeed, Figure 9.8 may be regarded as a T-Φ stability/instability phase diagram for this system. Below the boundary curve the dispersion is thermodynamically stable, but above it flocculation occurs, akin to phase separation, i.e., a floc phase separates and coexists with the dispersed phase (containing primarily singlet particles, as observed by a light scattering analysis). Moving horizontally (at constant temperature) across the phase diagram, the intersection with the boundary line corresponds to the c.f.Φ for that temperature.

The effect of free PEO on the stability of PS-PEO lattices has been studied by Cowell et al. [6], and Vincent et al. [9]. In these studies it was shown that at a given volume fraction of latex ($\Phi <$ c.f.Φ, for the electrolyte concentration and temperature, concerned), on increasing the concentration of added PEO, a critical concentration, c_p^+, was reached beyond which weak, reversible flocculation occurs, similar to that

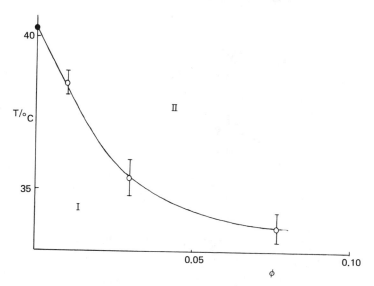

FIGURE 9.8   c.f.T. versus Φ plot for PS-PEO 2000 in 0.26 mol dm^{-3} MgSO$_4$ solution [21]. I indicates stability (single phase); II indicates instability (disperse phase + floc phase). ($\bullet$) Turbidity measurement; (o) viscometry measurement.

occurring when Φ > c.f.Φ. The extent of flocculation increases up to
some maximum value and then decreases again, until beyond a second
critical value of the free polymer concentration $c_p^{\ddagger}$, the dispersion becomes thermodynamically stable again, this stability continuing up to the
pure polymer (melt) situation. The c_p^+ and $c_p^{\ddagger}$ values are critically
dependent on the molecular weight of both the anchored PEO chains and
the added PEO. They also depend on Φ and indeed a three-component
(constant temperature) phase diagram may be constructed, delineating the
stability/instability regions, i.e., the single-phase and two-phase (floc +
dispersed) regions, as discussed above for two-component systems. The
three components here are PS-PEO particles + free PEO + solvent (e.g.,
MgSO$_4$ solution). A typical example is shown in Figure 9.9 for PS-PEO
750 latex plus PEO 10,000 (in water and 0.065 mol dm^{-3} MgSO$_4$ solution)
and PS-PEO-750 latex plus PEO 400 plus water. The instability regions
were determined from dlog τ/dlob λ, as described previously, over the Φ
range 10^{-5} to 10^{-3}. Measurements below Φ = 10^{-5} were not possible,
hence the broken lines in Figure 9.9 are guidelines rather than based on
experimental data. Clearly, however, as with PEO 400, there should
exist a critical value of Φ below which no flocculation is observed at any

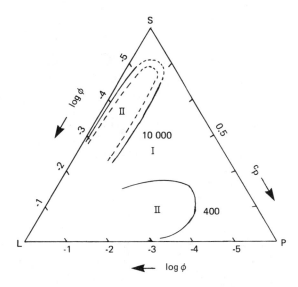

FIGURE 9.9   Three-component stability (region I)/instability (region II)
phase diagram for PS-PEO 750 latex (L), plus water (—)
or plus 0.065 mol dm^{-3} MgSO$_4$ (---) solution (S), plus
added PEO (P) of the molecular weight indicated. 25°C [9].

c_p. Since the depth of the minimum (G_{min}) in the interparticle interactions increases with increasing molecular weight of added PEO [9], one
can expect this critical value of Φ to decrease with increasing molecular
weight, in accord with equation (4) and also the experimental findings.
Although, as was discussed in the previous section, several theories [9,
15-17] have been proposed which purport to account for c_p^+, no convincing,
quantitative theory for $c_p^{\ddagger}$ as yet exists. Feigin and Napper claim to
account for $c_p^{\ddagger}$ in their theory [17] but they imply that the stability observed for $c_p > c_p^{\ddagger}$ is of the kinetic type, rather than true thermodynamic
stability. The experimental evidence is against this: the boundary line
(Figure 9.9) may be crossed reversibly in both directions at the same point
for $c_p^{\ddagger}$; furthermore, one would not expect to see such a critical Φ-
dependence of the boundary line anyway if the stabilization observed were
purely kinetic in type, i.e., corresponding to the presence of a significant
energy barrier in the interparticle interaction curve.

Figure 9.10 shows the three-component phase diagram for a PS-PEO
750 latex plus PEO 1000 and 10,000 plus 0.065 mol dm^{-3} MgSO$_4$ solution
at 48.5°C. The c.f.T. of this latex was found to be 48.5°C at $\Phi = 5 \times 10^{-4}$.
It may be seen that the two stability/instability boundary curves shown in
Figure 9.10 both intersect the latex (L) - solvent (S) axis at $\Phi \sim 5 \times 10^{-4}$;

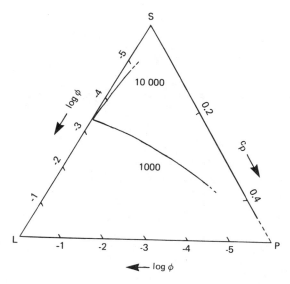

FIGURE 9.10   Three-component stability (region I)/instability (region II)
phase diagram for PS-PEO 750 latex (L), plus 0.065 mol
dm^{-3} MgSO$_4$ solution (S), plus PEO (P) of the molecular
weight indicated, 48.5°C [9].

this value of Φ, therefore, corresponds to the c.f. Φ for this latex in the absence of added PEO.

The experimental results presented in this section have all been for aqueous systems. Similar results could have been quoted for nonaqueous systems. By way of example Figure 9.11 shows one result for the three-component phase diagram [22] for the system PAN-PDMS latex, plus n-hexane plus added PPMS, where the PAN-PDMS latex is a polyacrylonitrile (PAN) latex particle, stabilized by polydimethylsiloxane (PDMS)-polystyrene-polydimethylsiloxane ABA block copolymer, and PPMS is poly(phenylmethylsiloxane) (number-average molecular weight, 1800. Here the actual experimental points are shown for c_p^+ and $c_p^{\ddagger}$. Note that, with respect to the $c_p^{\ddagger}$ points, some points were obtained on increasing c_p, while others were obtained on decreasing c_p i.e., by adding n-hexane to a dispersion of the latex in pure PPMS. Clearly Figure 9.11 shows essentially the same features as Figure 9.9 for the aqueous system. The T-Φ behavior for PAN-PDMS lattices in hydrocarbon fluids has been investigated by Everett and Stageman [23, 24] and again, the trends in behavior are similar to those observed with aqueous systems, except that because the temperature-composition polymer solution phase diagram for PDMS +

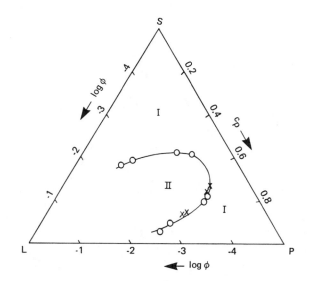

FIGURE 9.11   Three-component stability (region I)/instability (region II) phase diagram for PAN/PDMS latex (L), plus n-hexane (S), plus PPMS (molecular weight 1800) (P). O points obtained on increasing PPMS concentration; x points obtained on increasing hexane concentration [22].

n-hydrocarbons show both upper and lower critical solution (consolute) temperatures, so both upper and lower critical flocculation temperatures are observed.

CONCLUSIONS

This paper has attempted to illustrate the various, subtle ways in which the interplay of several effects can lead to weak, reversible flocculation in dispersions of sterically-stabilized effects. It has been shown that the two major parameters controlling the stability/instability behavior of systems of this type are the particle volume fraction (or number density) and the depth of the interaction free energy minimum (G_{min}). The latter is, in turn, largely determined by the thickness of the adsorbed layer, the polymer-solvent interaction parameter and the concentration of free polymer in solution.

REFERENCES

1. Vincent, B. Adv. Colloid Interface Sci., 4:193, 1974.
2. Vincent, B. and Whittington, S. Surface and Colloid Science (E. Matijević, ed.), New York: Plenum, 12:1, 1982.
3. See e.g. Ottewill R. H. Specialist Periodical Report (Chemical Soc. London), 1:183, 1973.
4. Smitham, J. B., Evans, R. and Napper, D. H. J. Chem. Soc. Faraday Trans. I, 71:285, 1975.
5. Napper, D. H. J. Colloid Interface Sci., 58:390, 1977.
6. Cowell, C., Li-In-On, R. and Vincent, B. J. Chem. Soc. Faraday Trans. I, 74:377, 1978.
7. Kitchener, J. A. Brit. Polymer J., 4:217, 1972.
8. Long, J. A., Osmond, D. W. J. and Vincent B. J. Colloid Interface Sci., 42:545, 1973.
9. Vincent, B., Luckham, P. F. and Waite, F. A. J. Colloid Interface Sci., 73:508, 1980.
10. Hamaker, H. C. Physica (Utrecht), 4:1058, 1937.
11. Vold, M. J. J. Colloid Sci., 16:1, 1961.
12. Vincent, B. J. Colloid Interface Sci., 42:270, 1973.
13. Croucher, M. D. and Hair, M. L. J. Phys. Chem., 83:1712, 1978.
14. Evans, R. and Napper, D. H. J. Chem. Soc. Faraday Trans. I, 73:1377, 1977.
15. Vrij, A. Pure App. Chem., 48:471, 1976.
16. Joanny, J. F., Leibler, L. and de Gennes, P. G. J. Polymer Sci. (Physics), 17:1073, 1979.
17. Feigin, R. I. and Napper, D. H. J. Colloid Interface Sci., 74:567; 75:525, 1980.

18. Lambe, R., Ph.D. Thesis (Bristol), 1979, to be published; see also reference 21.
19. Boucher, E. A. and Hines, P. D. J. Polymer Sci. (Physics), 14: 224, 1976.
20. Napper, D. H. J. Colloid Interface Sci., 29:168, 1969; 32:106, 1970.
21. Cowell, C. and Vincent, B., in Effect of Polymers on Dispersion Properties, Th. F. Tadros (ed.), New York: Academic Press, in p. 263, 1982.
22. Clarke, J., Ph.D. Thesis (Bristol), 1980.
23. Everett, D. H. and Stageman, J. F. Colloid Polymer Sci., 255:293, 1977.
24. Everett, D. H. and Stageman, J. F. Faraday Disc. Chem. Soc., 65:230, 1978.

SOME ASPECTS OF HIGHLY CONCENTRATED PIGMENT DISPERSIONS WITH MINIMIZED YIELD-VALUE FOR COATINGS APPLICATIONS

J. Kunnen and E. M. A. de Jong

Sikkens B. V.
Sassenheim, The Netherlands

INTRODUCTION

When trying to prepare highly concentrated dispersions of organic pigments in liquid media, the experimenter is often halted by the development of a yield value which strongly increases with concentration, especially in organic media. A dough-consistency is reached at a pigment volume concentration (PVC) far below the maximum that can be expected (full packing at about 70% volume).

The authors have noted that in some surfactant or resin solutions, however, very high PVC can be reached—a state which we will call "ultimate dispersion." These solutions differ from all other dispersing media in that they show gelling phenomena in certain concentration ranges. This observation suggested the possibility that the viscosity of an adsorbed layer at a pigment surface could be of importance for the stability of the dispersion in such a way that a solid immobile adsorbed layer would be a requirement for obtaining low-yield values. Thus, we assume that if we want to obtain an idea of what the viscosity could be in the adsorbed volume, we first look at the viscous properties of solution which has the same composition as the adsorbed layer. We further assume that the outer surface of an adsorbed layer will have an interfacial tension against the continuous phase.

Overbeek [1] pictures the case of the coalescence of two quartz particles covered with an adsorbed water layer and dispersed in oil without a stabilizer. He illustrates the decrease of the oil/water interface by coalescence of the water layers around the two particles.

Such a mechanism assumes the presence of an interfacial tension at the water/oil interface of the adsorbed layer. The adsorbed envelope must possess a certain mobility and deformability. In this discussion we will state reasons why we think that this mechanism is not confined to an immiscible adsorbed layer alone, but may be expected of all layers that are adsorbed on solid particles in a liquid medium.

It is this mechanism we would like to present as a probable cause of an early yield value at low PVC in the dispersion of organic pigments. Our hypothesis is that the slightest remnant of an interfacial tension, also in a not sharply-defined boundary, is capable of producing non-negligible forces of attraction on touching. These forces will have no effect when the adsorbed layer is gelled or solid. The purpose of this communication is to show and discuss three series of dispersions. In each series the degree of dispersion runs from "ultimate" to "unstable." Several changes in the properties of the continuous phases will be followed. In our view, these changes can be explained in such a way that they sypport our hypothesis.

The three series of organic pigment dispersions were prepared in a red devil. The amount of surfactant was constant and in large excess for each of the series. The maximum PVC obtained was taken as an indication of the stability of the dispersion. A high PVC means a high stability of the dispersion in this paper. We determined the excess adsorbed amount of the surfactants. The composition of the layer is calculated, wherever possible, by the Reháček method [2]. The total layer thickness was calculated from the composition. Gelling range and viscosity maxima were noted from viscosity measurements of the continuous phase at several surfactant concentrations. As a sign of the presence of micelles at higher concentrations, the Thomas factor is introduced and used. The change in cloud point emperatures is followed and the foam stability noted of the continuous phase.

We know that in trying to understand these phenomena we move completely out of the area of normal colloid chemical research. The studies involve (a) organic pigment surfaces on which no strong adsorption has been reported; (b) very high-volume concentrations; (c) high-surfactant concentration; (d) low molecular-weight surfactants; and (e) a definition of stability (low-yield value at high concentrations) that is not normally used.

MATERIALS

Hansagelb G01, Pigment yellow 1: chemically, a neutral acrylamide; obtained from Hoechst. Density 1.48; spec. surface 16 m^2/g or 23.7 m^2/cm^3.

Permanentrot FGR, Pigment red 112: chemically, an arylamide; obtained from Hoechst. Density 1.41; spec. surface 37 m^2/g or 52.2 m^2/cm^3.

Paliotolgelb 1770: chemically, an Azomethine Ni-complex; obtained from BASF. Density 1.62; spec. surface 14 m^2/g or 22.7 m^2/cm^3.

N-methyl-formamide 99% from Aldrich Europe: b.p. 180°-185°C. Mol. weight 59.07; dielectric constant 180.

Butyl-oxitol from Shell: Boiling range 167°-173°C.

1 Methyl-2-pyrrolidinone, obtained from GAF.

Berol 09 from SOAB: A nonylphenol with 9-10 ethyleneoxide groups; this surfactant is a liquid.

Berol 292 from SOAB Sweden: Nonylphenol with about 20 ethylenoxide groups. A waxy solid.

Shellsol T from Shell: A high boiling mixture of aliphatics.

Resin P. preparation: 1 mol. glycerol monosterate (Hymono 1103) was reacted with 1.5 mol. toluenediisocyanate in the presence of diazobicyclooctane (Dabco). Almost all NCO left was then reacted with soyamine (Armeen S) and the remainder with butanol. Excess butanol and solvent were distilled off under vacuum and replaced by Shellsol T.

METHODS AND DEFINITIONS

Preparation of the Dispersions

The components were premixed and then added to a 250 cm^3 metal container, filled with 150 g of 3 mm glass beads. Dispersion was carried out in a red devil. Every 30 minutes the containers were opened to add more pigment. This was continued until it was judged that any further addition of pigment would prevent good dispersion by either the viscosity or the yield value that had developed. All samples were dispersed for the same length of time.

Maximum Obtainable PVC

This maximum PVC is an arbitrary value dependent on the way the dispersion is made. The values we found vary over a wide range and

provide a good illustration of the relationship we want to show when we use
the red devil.

Definition of Adsorption

When in this paper the word adsorption is used, we will mean every
mechanism by which molecules are in the vicinity of the pigment surface,
move with it, and are centrifuged off with it. Such a mechanism may be,
for instance, "rejection adsorption" as mentioned by Osmond [3].

Calculating Composition and Layer Thickness
from the Adsorption Data

Two methods are available. In the first method, at two or more
concentrations of surfactant or resin, the excess adsorption is determined.
Following Reháček [2] a straight line through these points in an adsorption/
composition diagram will give the composition of the adsorbed layer at the
point of zero-excess adsorption. As the excess adsorption values have a
standard deviation this leads to a range of values for zero excess composition. From the material balance before and after adsorption, the total
adsorbed volume of resin + solvents can be calculated. This volume
divided by the pigment surface present in the dispersion gives the thickness
of the absorbed layer. Reháček's method is only applicable when the adsorption is independent of the surfactant concentration and because the
solvent composition is not changed in these measurements, we will consider this as approximately correct.

The second method works with adsorption at two different PVC values.
The composition of the dispersing medium is kept constant. Two values
for the equilibrium composition of the continuous phase (after adsorption)
will be entered into two equations from which volume and composition of
the adsorbed layer can be calculated. The method requires a very small
standard of deviation for the composition.

Viscosity Measurement and Yield Values

Such measurements were carried out with a Rotovisko R V2 at 23°C.
Liquids were measured up to 4000 sec^{-1} and pigment dispersions to 90
sec^{-1}. In all cases rheograms were obtained. The yield value was taken
at the position where the curve leaves the shear stress axis. Viscosities
were calculated from the slope of the tangent at the highest shear rate.

Excess Adsorption

Excess adsorption is calculated from the difference in the concentrations of the surfactants in the continuous phase, before and after adsorption. Solvents were removed by evaporation at 150°C for 3 h. Preliminary determination showed that reproducible values were obtained that way. The pigments were removed by centrifugation in an Hereaus Labofuge 15000 at 13.500 rev/min for 15 min. Tubes were closed to prevent evaporation. All concentrations and adsorptions were determined as an average of 5 measurements.

The Thomas Factor f_{Th}

We have found [4] that the Thomas equation [5], which combines measuring data of 16 different authors on the viscosity of solid spherical particle dispersions to high concentrations, can be used far outside its original area of application. By the introduction of a volume factor f_{Th}, we obtain

$$\eta_R = 1 + 2.5\,f_{Th}\,\phi + 10.06\,f_{Th}^2\,\phi^2 + 0.00273\,\exp(16.6\,f_{Th}\phi) \qquad (1)$$

where ϕ is a volume fraction. For $f_{Th} = 1$, we have solid spherical particles; $f_{Th} < 1$ indicates liquid-like behavior and $f_{Th} > 1$ indicates solid nonspherical or swollen spherical particles.

The formula is exact for certain dispersions and solutions. It is qualitative and easy to apply for most dispersions and solutions when a table of its values is available. The product $f_{Th}\,\phi$ is the "equivalent solid-spherical particle volume" according to Thomas. It gives an impression of the way dissolved molecules and dispersed particles are present in solution, especially at high concentrations.

Foam Stability

Foam stability was measured as the time in which a volume of foam, produced by shaking for 30 sec, was reduced to half its original height.

Cloud-Point Temperatures

Cloud-point temperatures were determined with DIN 53817 as a guide. When these temperatures rose above 100°C, 10% by weight NaCl was added to the solvent mixture. An increasing cloud point temperature is generally connected with a higher CMC. Measurements by Becher [6] suggest that micelle formation will no longer occur when the concentration of added ethanol or dioxane reaches approximately 25% to 30% by weight of the aqueous solution. We think that the increasing cloudpoint temperatures in our experiments can also be explained as the gradual disappearance of micelles. In this way they support the conclusions drawn from the change of f_{Th}.

RESULTS

The preparation of the dispersions and further measurements were carried out in the order that is indicated in the introduction. The results of the experiments, measurements and calculations on the three systems are given in Tables 10.1, 10.2 and 10.3. When, at the surfactant concentrations used in the experiments, the amount of adsorption indicated that no saturation was yet reached, no figure was entered under composition of absorbed layer and layer thickness.

Some general conclusions can be drawn before we discuss these results further. The three systems show that strong positive adsorption of surfactant corresponds closely with the presence of micelles or aggregates of the surfactants. The figures show that it is quite likely that these pigments in these media are always fully covered with something. Adsorption of solvent alone may result in thick layers. Maximum PVC is reached when gelling is possible in the corresponding solvent mixture.

DISCUSSION

In the three series presented, we have moved from a highly-stabilized situation to one of lower stability as the maximum attainable PVC shows. Systems A and B start in water. As soon as water is used the possibility of coulombic stabilization must be considered. In system A, and also in C, we have purposely chosen for the admixture of a solvent with a higher dielectric constant than the liquid in the most stable state. In this way we avoid the remark that in two of our cases the change to lower stability is caused by the fact that coulombic repulsion is not effective at lower dielectric constant.

At present, the theory of coulombic stabilization of highly concentrated dispersions, with adsorbed layers, in highly conductive media with high dielectric constant, is not developed enough to answer the question of whether or not stabilization by coulombic repulsion is important. At the same time it cannot be ruled out. We must leave room for the possibility that it still plays a part.

An easy explanation of these flocculation phenomena seems possible in the language of steric repulsion when we close our eyes for all the other information we have presented and only look at the maximum PVC and the

TABLE 10.1

System A—Results of Measurements and Calculations on
Dispersions of Hansayellow G 01 in Mixtures of Water
and N-methylformamide, with Berol 292 as Surfactant

(Surfactant concentration 8% vol. of sum of surfactant + liquids)

Composition of solvent mixture (volume)	H_2O	1	3	1	1	–
	NMF	–	1	1	3	1
Maximum PVC in red devil		42	39	39	23	21
Excess ads. in mg surfactant/cm^3 pigment	at 4% surf.	26.5	27.0	-1.31	-22.4	-72.5
	at 8% surf.	36.7	19.3	-17.1	-32.3	-130.8
Composition adsorbed layer in % surfactant		(35)[*]	17.9	3.5	0	0.2
Uncertainty range in %		38-100	13.5-30	4.1-2.6	1.8-0	1.5-0
Layer thickness (nm) at 4% surf.		(3)[*]	7.7	12.3	21.5	66
at 8% surf.		(4.7)[*]	6.7	12.2	18.4	67
Gelling range		40-80	40	–	–	–
Viscosity maximum		yes	yes	yes	no	no
f_{Th} at 10% surfactant		2.75	2.22	1.91	1.33	1.30
Foam stability in seconds		2400	360	180	20	5
Cloud point temp. at 10% NaCl in °C		71	95	>110	–	–

[*]Estimated

TABLE 10.2

System B—Results of Measurements and Calculations on
Dispersions of Permanent Rot in Mixtures of Water
and 1-Methyl-2-pyrrolidinone with Berol 09 as Surfactant

(Surfactant concentration: 10% vol.)

Composition of solvent mixture (volume)						
	H$_2$O	1	3	1	1	–
	NMP	–	1	1	3	1
Maximum PVC in red devil		36	36	31	26	16
Excess ads. in mg surfactant/cm^3 pigment	at 5% surf.	58.1	21.0	–6.8	–67.6	–57
	at 10% surf.	39.6	20.6	–6.2	–	–72
Composition adsorbed layer in % surfactant		20.2	>10	<10	0	0
Layer thickness (nm) at 5% surf.		6.0	–	–	23.1	17
at 10% surf.		6.0	–	–	–	15
Gelling range		40–70	–	–	–	–
Viscosity maximum		yes	yes	–	no	no
f_{Th} at 10% surfactant		3.93	1.47	1.44	1.00	0.90
Foam stability in seconds		18000	18000	7200	110	3
Cloud point temp. at 10% NaCl in °C		53	78	>107	–	–

TABLE 10.3
System C—Results of Measurements and Calculations on
Dispersions of Paliotolgelb 1770 in Mixtures of
Butyl-oxitol Shellsol T, with Resin P as Surfactant.

(Surfactant concentration 9.11% vol. of surfactant + solvent.)

Composition of solvent mixture (volume)	A*	1	3	1	1	–
	B*	–	1	1	3	1
Maximum PVC in red devil		43	35	31	29	26
Excess ads. in mg surfactant/cm³ pigment	at 4% surf.	51.0	35.4	18.5	11.2	-7.3
	at 8% surf.	52.5	40.8	24.3	6.7	-5.8
Composition adsorbed layer in % surfactant		–	–	–	15.4	0
Layer thickness (nm) at 4% surf.		–	–	–	4.8	6.8
at 8% surf.		–	–	–	4.8	2.9
Gelling range		30-70	–	–	–	–
Viscosity maximum		yes	yes	no	no	no
f_{Th} at 10% surfactant		2.91	1.85	1.38	1.20	1.21

*A = Shellsol T; B = Butyloxytol.

corresponding amount of surfactant/resin that is adsorbed. It says that
the more there is adsorbed, the more stable the dispersion, a seemingly
simple case. At the negative excess adsorption the surface is bare and
London-van der Waals forces can freely develop, resulting in a high-yield
value. It is true that the length of the tails in the two Berol surfactants
are not very impressive as a cause of steric reuplsion, but the fatty urea
in series C with its average molecular weight 2000 seems a flawless
example.

When we now allow the other information to be taken into consideration, then from system A, B, andC, we can remark that it is not correct
to state that the adsorption changes from full coverage to no coverage. As
far as thickness could be determined, it is much more likely that over the
entire range, each pigment is covered with something, and in several
cases more than fully covered. When about 4 nm thickness in system A is
capable of preventing particle approach to a measurable London-van de
Waals forces level, why is a 15 x thicker layer in this series not? This
layer is so strongly attached to the pigment surface that it cannot be
centrifuged off and is not destroyed at 13.000 rev/minute. It is not likely
that the adsorbed layer itself, having almost the same density and composition as the continuous phase, develops an extra London-van der Waals
attracting force to account for the yield value.

The strong correlation between a high Thomas factor, f_{Th}, and the
Max. PVC, means that in these systems with comparatively low-energy
pigment surfaces, the resins or surfactant should be present in the
continuous phase as aggregates or micelles to promote adsorption. This
is an indication that the surfactant is not in an ideal surrounding and has
the tendency to move to any interphase that is present. Lipatov [7]
developed such ideas, according to which macromolecular aggregates,
and not single macromolecules, pass to the surface in the case of polymer
adsorption. Such an approach leads to the investigation of the structure
of the adsorbed film, from the point of view of structure formation in the
polymer solution.

In our view, a high Thomas factor is not enough for ultimate stabilization. On increasing the concentration of the surfactant f_{Th} should change
so little that the product $f_{Th} \phi$ reaches the value 0.75. This is a gel value.

Our hypothesis on the causes of stability of organic pigment dispersion, as described above, is in accordance with our experience that ultimate dispersion is possible only when the continuous phase has gelling
properties somewhere at higher surfactant concentrations. Adsorbed
layers, which we see as rearranged micelles on the surface, with a resin/
surfactant concentration that is not too far removed from gelling concentrations, will gel under the immobilizing adsorption forces that press the
molecules to thepigment interface. We further assume that most adsorbed
layers will have an interfacial tension. This tension will produce attrac-

tive forces between particles in the way Overbeek has pictured, when the adsorbed layer is liquidlike, but no attraction if it is gelled or solid. At the same time we expect that the other possible attracting and repulsive forces will also be present, but we have the impression that the cause of instability that we just have described, dominates in the practical field of making pigment concentrates or pigment pastes.

The idea is not new; Russian workers, Rehbinder and Taubmann [8] reported that it is generally acknowledged that concentrated suspensions require the existence of a mechanical (viscous or elastic) barrier to impede the approach of particles if they are to show maximum stability. Sonntag and Strenge [9] devoted a paragraph to the gel-like properties of protecting colloids and the mechanical barrier of adsorbed layers.

As to the interfacial tension of adsorbed layers, it was already mentioned earlier that its existence seems quite normal and acceptable as a cause of particle attraction on contact, if the adsorbed layer is immiscible with the continuous phase. It comes in addition to the repulsive coulombic force and the attractive London-van der Waals force. If for any reason in the Overbeek picture the adsorbed layer would have become solid or highly viscous, such an attractive force could not develop.

In our view, it is strange that for all other cases where the adsorbed layer grows from the continuous phase or where molecules are grafted on the pigment particle surface, we have not found one single word or discussion relating to the question if such an interfacial tension either exists or not in such cases. Still it is important, because the smallest interfacial tension is capable of producing attracting forces that cannot be neglected.

The interfacial tension between touching volumes of different composition is such a general case that its total absence needs extensive proof. Interfacial tension is normal between the two phases of partly miscible liquids. It has been measured in micelles. Its presence is used to explain the limitation in growth of micelles by the counteracting increase of surface energy. Micelles are adsorbed from the continuous phase. When they are rearranged at the surface, will they lose their interfacial tension?

The presence of an interfacial tension at the inner and outer surfaces of an adsorbed layer, in the case of emulsions, seems quite normal and acceptable [10, 11]. It is very unlikely that when the emulsion droplet becomes highly viscous or solid, the interfacial tension at the outer surface will suddenly disappear.

The stability of foams on the water-containing media corresponds closely with the dispersion properties of such media for organic pigments, as the figures in system A and B show. In a stable foam the continuous phase is showing what it is capable of at inert (air) surfaces in the way of producing high viscosity interfaces. Long lasting foam is connected with high viscosity [12].

The peculiar properties of both NMP and NMF adsorbed in multi-layers, reminds us of "structured" layers of n-Alkane at a Graphon surface by Parfitt and Tideswell [13].

The increase of yield value with concentration is in the stable range quite different from that in the unstable range. The yield values follow an exponential law, as was found by Weltmann [14]:

$$\tau = \tau_0 \exp{(\alpha \phi)} \tag{2}$$

in which τ_0 is the yield value at zero concentration and α is a constant (Table 10.4).

As a consequence of our hypothesis, we may expect that when a sufficient amount of resin is adsorbed on an organic pigment to completely cover its surface, this will not produce stability unless the resin solution has gelling properties. In the past years, we have collected a multitude of examples of oil and water soluble alkyds that show good adsorption. They do not gel in solution and nowhere was ultimate dispersion obtained.

In this study, we have purposely excluded inorganic pigments. Inorganic pigments generally possess strongly adsorbing surfaces. So strongly adsorbing that, irrespective of the way the surfactants are in solution or whether the continuous phase has gelling properties or not, a solid-like adsorbed layer can still be produced, resulting in ultimate dispersion.

TABLE 10.4
Yield Values in System A

Solvent composition	PVC	Yield value in Pa
NMF : H_2O	20	386
3 : 1 by vol	15	111
	10	39
100% H_2O	50	16.5
	40	7.6
	30	3.1

REFERENCES

1. Overbeek, J. Th. G. In "Colloid Science," H. R. Kruyt, ed., p. 358, 1952.
2. Rehácek, K. Farbe und Lack, 76:656, 1970.
3. Osmond, D. W. J. Proc. Chem. Inst. Canada, Flocculation and Dispersion Symposium, Toronto, November 1974.
4. de Jong, E. M. A. and Kunnen, J. Euromech Colloquium 104, "Mechanics of colloidal dispersions," Leuven, Sept. 1978.
5. Thomas, J. G. J. Colloid Sci., 20:267, 1965.
6. Becher, P. J. Colloid Sci., 20:728, 1965.
7. Lipatov, Yu. S. and Sergeeva, L. M. "Adsorption of polymers," New York: John Wiley & Sons, p. 73, 1974.
8. Rehbinder, P. A. and Taubman, A. B. Colloid J. USSR 23:301, 1961.
9. Sonntag, H. and Strenge, K. "Koagulation und Stabilität disperser Systeme." VEB Deutscher Verlag der Wissenschaften, Berlin, p. 137, 1970.
10. Prince, L. N. "Microemulsions," New York: Academic Press, 1977.
11. Lopulissa, J. S., Mellema, J. and Oosterbroek, M. Proc. 8th Int. Congress on Rheology, 1980.
12. Bikerman, J. J. Foams, New York: Springer Verlag,Heidelberg, p. 173, 1973.
13. Parfitt, G. D. and Tideswell, M. W. J. Colloid Interface Sci., 79: 518, 1981.
14. Weltmann, R. N. Rheology, Vol. 3, Ed. F. R. Eirich, p. 222, 1965.

SAG CONTROL AGENTS FOR RHEOLOGY CONTROL IN AUTOMOTIVE TOPCOATS

A. Heeringa and P. J. G. van Hensbergen

Automotive Research Department
Sikkens B. V.
Sassenheim, The Netherlands

INTRODUCTION

In the initial stages of a search for low molecular-weight resins for high-gloss thermosetting automotive topcoats, the need for a special rheology control emerged. The conventional products like silicas and bentones were less suitable because excessive amounts were required to prevent sagging. The literature indicates that a solution might possibly be offered by application of distinct classes of urethanes [1, 2] and ureas [3-6].
For automotive topcoats the combination of good levelling and flow with an acceptable storage stability of the paint is required. A good compromise was found in a series of diureas—the adducts of hexamethylene diisocyanate with primary monoamines of a different chemical nature. For reasons of convenience these adducts were prepared "in situ" in the plasticizing part of the binder. This kind of antisagging agent fitted very well into the concept of sag control agents.

CONCEPT OF SAG CONTROL AGENTS

The rheology control desired should cause a high viscosity at low-shear rates (antisagging) and a low contribution to the viscosity at high shear rates (application). For wide applicability, this should preferably

be achieved by an additive; a Sag Control Agent (SCA). For the high
standards of appearance and performance for automotive topcoats there
are requirements

1. adequate rheology control—good sag control together with
    good levelling.
2. complete disappearance of the SCA as a physical identity
    in the process of curing (gloss).
3. acceptable storage stability of the paint.
4. no significant decrease in solids content.
5. no changes in overall coating properties.
6. reproducible preparation of the SCA.

These requirements have to be translated into physical and chemical
terms. When prepared "in situ" in resins the adducts of hexamethylene
diisocyanate (HDI) and primary monoamines showed promise as SCAs. The
adducts were present as finely dispersed elongated particles as can be
seen in Figure 11.1 showing a Scanning Electron Microscope (SEM) picture
of an adduct of HDI and Benzylamine (BA) as prepared in and isolated from
an acrylic resin.

The rheological properties of the resin containing the SCA were in
line with what was sought, as illustrated in Figure 11.2. All the scans of
shear rate versus stress were run with the Haake Rotovisco RV_2, 50SVI
at $23°C$. Program 5'-1'-5' over a range of D=0 to D=45.6 s^{-1}. The
apparent yield value was determined by extrapolation of the linear region
to the stress axis. This afforded at the same time the viscosity at
D=45.6 s^{-1}.

To get an insight into the behavior of this kind of SCA the initial
explorative work was restricted to the adducts of HDI and BA and cyclo-
hexylamine (CHA) in an automotive acrylic resin.

Acrylic resin:
$\overline{Mn}$ ~ 6000
OH number   ~ 60 (solids)
Acid number   5-8 (solids)
(51% by weight solids in a mixture of butanol and xylene)

The resin in all cases was modified with 5% by weight of the SCA on resin
solids according to a procedure for ambient temperature (see Appendix).
Any deviations from this procedure are given in the text together with the
relevant results of the experiments. The temperature rise following the
addition of the diisocyanate never exceeded $10°C$. The SCAs were isolated
for SEM photographs by diluting the SCA containing resin 500 times with

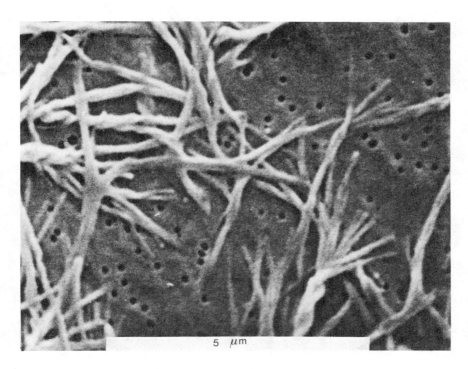

FIGURE 11.1   SEM picture of HDI-BA as prepared in an acrylic resin.

xylene and filtering through a micropore filter.   The SCA was then washed
and dried.

PHYSICAL SYSTEM

In accordance with what Figures 11.1 and 11.2 suggest, it was
assumed that the antisagging properties were the result of the SCAs forming a network of partly flocculated elongated particles in the liquid paint.
Shear forces above a certain magnitude would bring a reordering of the
network to give the desired shear thinning during application (spraying and
levelling).   In this picture the degree of flocculation, the strength of the
network also depends on the adsorbed layer on the surface of the particles.
In this sense the medium of preparation (resin) was also thought to play
a role in the efficacy of the SCA.   The two SCAs as prepared in the
acrylic resin did not show any sign of flocculation, such as formation of
dense agglomerates.   This was not even apparent at infinite dilutions in

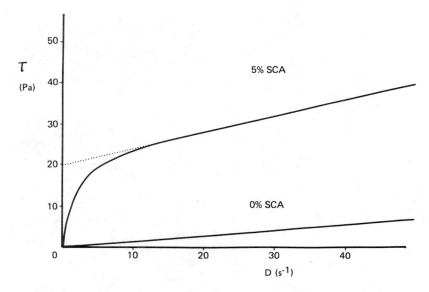

FIGURE 11.2   Scan of shear rate D versus stress τ of an acrylic resin as
              such and containing 5% by weight HDI-BA on dry resin
              solids.

xylene, acetone or paraffin. Preparations in these liquids gave coarser
materials with essentially the same behavior pattern. This suggests that
there was a steric barrier preventing the fiberlike particles aligning.

     A model accommodating all these suggestions is that of the particles
being of an open tertiary structure as illustrated in Figure 11.3. These
structures are supposed to incorporate polymer and/or solvent. This
model accounts for the fact that these SCAs could not be separated from
the resins by centrifugation (here only 5 times enrichment after 1 h at
1500 rpm). This model also accounts for the fact that these SCAs were
present in all kinds of bent and twisted long shaped particles as shown in
Figure 11.1.

CHEMICAL SYSTEM

     The formation of the diurea compounds studied proceeded very fast
according to:

$$O{=}C{-}N{-}(CH_2)_6 - N{=}C{=}O + 2\ R{-}NH_2 \longrightarrow R{-}NH{-}\overset{O}{\overset{\|}{C}}{-}NH{-}(CH_2)_6{-}NH{-}\overset{O}{\overset{\|}{C}}{-}NH{-}R$$

HDI   BA(CHA)   HDI – BA(CHA)

FIGURE 11.3   Proposed structure of SCA particles.

Under the conditions applied side reaction, like possible urethane formation with the hydroxyl functions of the resins and solvents, played no substantial role. With a virtually quantitative isolation of the SCAs no
urethane bonds could be detected in these materials by IR. The regular
form of the particles found in the SEM pictures (Figure 11.1) strongly
suggest that the reaction products of HDI with BA and CHA were of
crystalline nature. This was confirmed by X-ray diffraction. So the
primary particles of the structures in the model (Figure 11.3) consisted
of crystalline needles. The crystallization in very fine needles was
assumed to be a consequence of the specific chemical nature of the diureas.
From their nature these diureas should show a strong tendency for hydrogen bonding via the urea bonds.

$$- N - C - NH -$$
$$- N - C - NH -$$

The association via the urea bonds would cause the crystalline
particles to grow with the highest speed in the direction perpendicular to
the urea bonds and then very strongly favored in the plane of hydrogen
bonding. Imperfections in the process of building the molecules into the
crystal lattice would not change the direction of the highest speed of
growth (Figure 11.4). However, the uncompensated urea bonds are
assumed to act in some cases as nuclei for a new direction of high speed
growth at a small angle with the original direction. This would then result
in branched fiberlike structures similar to those as shown in Figure 11.1.
This model of crystallization (low-lattice enthalpy because of the very
efficient hydrogen bonding) accounts for the fact that this kind of SCA showed
very low solubility in the common paint binders and solvents.

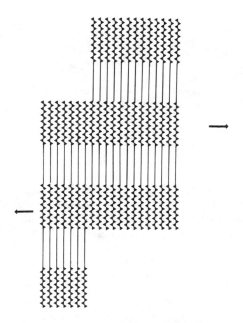

FIGURE 11.4   Arrangement of diurea molecules in crystals.

PARTICLE FORM AND SIZE

Since the SCAs were formed by a process of crystallization the parameters mentioned below would influence the form and size of the SCA particles.

Chemical Nature

To illustrate the dependence of the particle form and size on the chemical nature of the SCA the SEM pictures of the adducts HDI – BA and HDI – CHA are given in Figures 11.5 and 11.6, respectively. Both addition products were prepared under exactly the same conditions according to the procedure in the Appendix with instantaneous addition of HDI. The rheological data refer to the resin plus SCA with a resin solids content of 40% by weight and 5% by weight SCA on resin solids.

Temperature during Formation

To show the effect of temperature, SCAs were prepared at 70°C, ambient temperature and 0°C. The preparation procedure was essentially

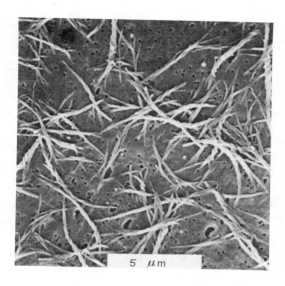

FIGURE 11.5   HDI–BA addition product. Resin with SCA at 23°C;
viscosity (D=45.6 s^{-1}) 38 cPa.s, apparent yield value
20 Pa.

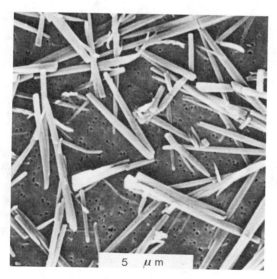

FIGURE 11.6   HDI–CHA addition product. Resin with SCA at 23°C;
viscosity (D=45.6 s^{-1}) 13 cPa.s, apparent yield value
0 Pa.

the same as described in the Appendix, HDI being added instantaneously at the different temperatures. The starting viscosities of the acrylic resin before addition of HDI were kept the same at the different temperatures by adjusting the solids content with xylene. The SEM pictures of the isolated SCAs are given in Figures 11.7, 11.8, and 11.9. The rheological data refer to the resin plus SCA adjusted with xylene to a resin solids content of 35% by weight and 5% SCA on resin solids.

Rate of Formation

The influence of the rate of addition of HDI during preparation on the particle form and size is illustrated in Figure 11.10 for the HDI – BA type SCA (procedure as in Appendix). The results for the instantaneous addition of HDI during the preparation of HDI – BA were given above (Figure 11.5).

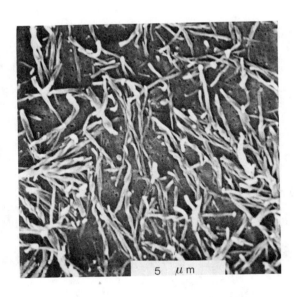

5 μ m

FIGURE 11.7   HDI-BA. Starting solids of acrylic resin 51% by weight. 70°C. Resin with SCA at 23°C; viscosity (D=45.6 s^{-1}) 18 cPa.s, apparent yield value 10 Pa.

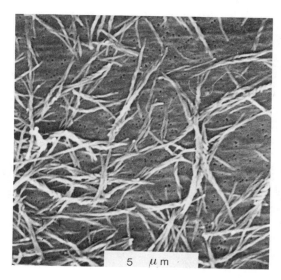

FIGURE 11.8   HDI–BA. Starting solids of acrylic resin 40% by weight.
Ambient temperature. Resin with SCA at 23°C; viscosity
(D=45.6 s^{-1}) 21 cPa.s, apparent yield value 15 Pa.

FIGURE 11.9   HDI–BA. Starting solids of acrylic resin 35% by weight.
0°C. Resin with SCA of 23°C; viscosity (D=45.6 s^{-1})
22 cPa.s, apparent yield value 17 Pa.

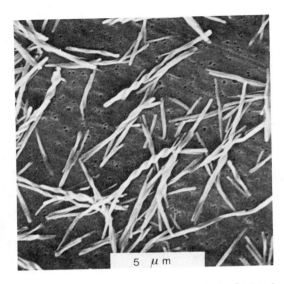

FIGURE 11.10   HDI-BA。 HDI added over a period of 10 min. Resin with
SCA at 23°C; viscosity (D=45.6 s^{-1}) 28 cPa.s, apparent
yield value 9 Pa.

Shear During Formation

Experiments with High-Solid Polyesters for stoving enamels showed
that the higher the shear rate the finer the isolated SCA particles. Varying
the stirrer speed at ambient temperature in the simple experimental set
up described in the Appendix however, did not change the size and form of
the SCA particles.

The conclusion from the above experiments is that the experimental
conditions must be carefully controlled for reproducible preparation of
SCAs.

REACTIVITY

A good SCA should be stable during storage and for high gloss automotive topcoats, completely disappear in the curing cycle. Chemically,
this means that during storage the SCA should be virtually insoluble in the
medium and during curing should react with the medium destroying the
strong hydrogen bonding in the crystal lattice.

Solubility

The SCA HDI-BA at ambient temperature could be recovered almost quantitatively from the standard acrylic resin and from several high-solid polyester resins, even when the resins were diluted with xylene or acetone to mixtures containing not more than $10^{-3}\%$ weight of the SCA. In these cases the isolated materials showed no traces of ester or urethane groups (IR). HDI-BA could be dissolved in the medium by adding large quantities of para toluene sulphonic acid and heating to $80°-100°C$. On cooling the SCA appeared again (fine crystalline needles) chemically unchanged (IR).

Reaction with the Medium

The SCAs (diureas) reacted with melamine crosslinkers under the conditions of stoving automotive thermosetting topcoats. The urea bond was involved in such a way that hydrogen bonding became impossible. The reactions occurring were identified by model experiments. The urea bonds were methylolated disrupting the hydrogen bonding and destroying the crystal structure. These reactions were catalyzed by acids [7].

The above findings meant that when working with blocked acid catalysts (high-solid systems) the storage stability tests should be run at the relevant temperatures ($20°-40°C$) because the results of the experiments run at higher temperatures ($60°-90°C$) could not easily be related to those of the lower temperatures.

SAG CONTROL AGENTS FOR AUTOMOTIVE TOPCOATS

The concept of Sag Control Agents incorporated the idea of application in high gloss thermosetting, melamine crosslinked, automotive topcoats [8]. So they could possibly solve the severe sagging problems with High-Solid (HS) automotive topcoats.

Since only a polyester HS topcoat was available this was taken as a model system to evaluate the potency of SCAs under extreme conditions. A prerequisite was good levelling, flow, and complete physical disappearance of the SCA in the curing cycle. In addition, the paints should have acceptable storage stability. The mechanical and chemical film properties had to remain within acceptable limits, so the SCAs could only be applied in small amounts.

The preliminary experiments, described above, suggested that the most promising way of adapting the SCA to a particular paint was by changing the chemistry. This meant the evaluation of a series of adducts of hexamethylene diisocyanate (HDI) and primary monoamines of a different

chemical nature all prepared "in situ" under exactly the same conditions.
When necessary the SCA properties could be optimized by changing the
conditions of preparation.

HIGH-SOLID POLYESTER AUTOMOTIVE TOPCOATS

For HS automotive topcoats, the conventional antisagging agents
(silicas, bentones, polyamides, etc.) were not suitable because excessive
amounts had to be added to prevent sagging on vertical surfaces. These
large amounts resulted in an unacceptable loss of gloss. SCAs were
supposed to afford a solution here. This is schematically illustrated in
Figure 11.11.
For the purpose a series of adducts of HDI with different primary
monoamines were evaluated in a HS automotive polyester topcoat.

    Polyester resin
    $\overline{Mn}$ ~550
    OH number   ~200
    COOH number ~ 3
    (Dissolved 80% by weight in a mixture of ethylglycolacetate and
    xylene (weight ratio 1:1))

The preparations were carried out at ambient temperature according to a
procedure very similar to that described for an acrylic resin in the

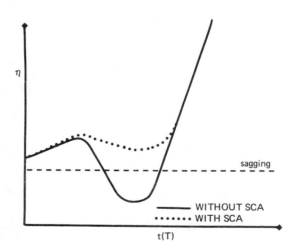

FIGURE 11.11   Schematic representation of the effect of a SCA on the
viscosity of a HS thermosetting topcoat on baking.

Appendix. The amount of SCA was 4% by weight on resin solids in all cases.
Table 11.1 summarizes the results.

SEM pictures of the SCAs (isolated as described above) gave very
similar results to that shown in Figure 11.1, except HDI-CHA and HDI-StA.
To illustrate the similarities and differences the SEM pictures of HDI-BA,
HDI-CHA, HDI-HeA and HDI-StA are given in Figure 11.2.

ANTISAGGING PROPERTIES

The antisagging properties of the SCA containing HS polyester resins
were measured in a white pigmented (p/b ratio 0.67) HS polyester automotive topcoat formulation. The formulations contained 3% by weight of SCA
on total binder. In a normal automotive curing cycle these formulations
needed a catalyst, which had to be blocked to arrive at a satisfactory
storage stability. Films having a wet layer thickness of respectively about
65 μm, 75 μm, and 85 μm were drawn with a doctor blade.
After a 4 min flashoff (panels in horizontal position) the films were
cured with the panels in the verticle position—curing cycle 25 min at 130°C.
After curing the layer thickness on each panel 1 cm below the top of the
film was measured at 3 places (about 1 cm apart). The experiments were
duplicated and the mean value of the 18 determinations was taken as a
m e asure for the SCA efficacy in this system. For DOI and gloss measurements panels were sprayed (30" DIN cup 4) and cured (25 min, 130°C)
horizontally. The results were summarized in Table 11.2, which clearly
indicates that HDI-BA gave the best compromise between solids content,
antisagging, properties, DOI, and gloss.
Extensive spraying tests with perforated panels showed that with HDI-
BA dry layer thicknesses of around 60 μm could be applied without running
into problems of sagging from verticle surfaces (as compared to without
SCA about 30 μm). With HDI-BuA and HDI-HeA this layer thickness was
about 50 μm. With HDI-CHA and HDI-StA the gloss had suffered. Using
unpigmented films, it was shown that these SCAs did not completely
disappear physically under the applied curing conditions.
To evaluate the possible role of the pigment in building up the antisagging properties Sag Balance [9] experiments were run with the white
pigmented and unpigmented paints containing 3% and 1.5% by weight HDI-BA
on binder solids. The paints were adjusted to a viscosity of 40" DIN cup 6.
Figure 11.13 shows that the presence of the TiO_2 pigment had no substantial
influence on the antisagging properties of the SCA HDI-BA in this HS polyester automotive topcoat formulation. The solids content of the coatings at
spray viscosity decreased when SCA was used (Table 11.3).
Casson plots [10] in the range of D(6000 - 90 s^{-1}) at 23°C indicate
that at infinite shear (at spray application D(10.000 - 40.000 s^{-1})) this
decrease was less than when measured at the viscosity of 30" DIN cup 4 and
23°C. This is illustrated in Figure 11.14.

TABLE 11.1

SCAs of HDI and Different Primary Monoamines as Prepared in a HS Polyester Resin

(80% by Weight in Ethyl Glycol Acetate and Xylene 1:1)

	HDI – Monoamine adduct			SCA in HS polyester resin 4% by weight on resin solids		
Code	Primary amine	M.P.[a] (°C)	Yield[b] SCA (%)	Fineness Hegman gauge (μm)	Apparent[c] yield value at 23°C (Pa)	Viscosity[c] $D=45.6\ s^{-1}$ $T=23°C$ (cPa. s)
HDI–BA	Benzylamine	221–222	96.4	< 5	22	114
HDI–CHA	Cyclohexylamine	229–230	90.3	13	2	72
HDI–PrA	n–Propylamine	202.5–203	78.9	< 5	9	110
HDI–BuA	n–Butylamine	199	88.9	< 5	22	126
HDI–AmA	n–Amylamine	187–188	86.0	< 5	21	93
HDI–HeA	n–Hexylamine	185	85.1	< 5	23	113
HDI–OcA	n–Octylamine	180–181	88.6	< 5	12	88
HDI–StA	Stearylamine	174	96.1	< 5	3	88
–	–	–	0	–	2	69

[a] SCAs prepared separately in acetone.

[b] Gravimetric determination by diluting the resin with acetone (500X), filtering, washing with acetone and drying.

[c] Both measured on return scan.

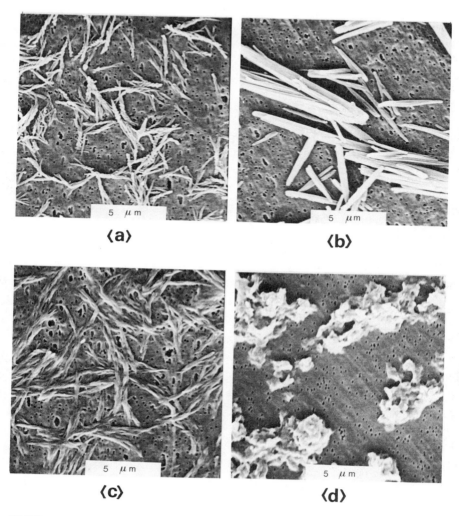

FIGURE 11.12   SEM pictures of (a) HDI-BA, (b) HDI-CHA, (c) HDI-HeA
and (d) HDI-StA as prepared in a HS polyester resin.

TABLE 11.2

Paint and Film Properties of White Pigmented HS Polyester
Automotive Topcoats with 3% SCA on Solid Binder

SCA (See Table 11.1)	Weight solids content at 30" DIN cup 4 after 1 hr 130°C (%)	Sagging panels cured vertically: Average layer thickness 1 cm below top of the film after cure (μm)	Gloss panels cured horizontally			
			DOI[a]	Gloss[b] 60°	20°	Dry layer thickness (μm)
HDI – BA	68.2	51.0	10	96	89	38 – 48
HDI – CHA	71.6	30.3	10	95	84	42 – 48
HDI – PrA	67.9	43.7	10	96	89	46 – 55
HDI – BuA	62.9	45.3	10	95	88	40 – 51
HDI – AmA	59.6	41.7	10	95	87	46 – 49
HDI – HeA	59.2	45.3	10	95	87	40 – 49
HDI – OcA	67.7	44.3	10	96	88	52 – 58
HDI – StA	70.2	37.7	9	94	81	47 – 52
–	72.1	30.1	10	95	88	40 – 43

[a] Distinctness-of-image: image clarity meter according to Ford
[b] Average of 3 determinations on one panel

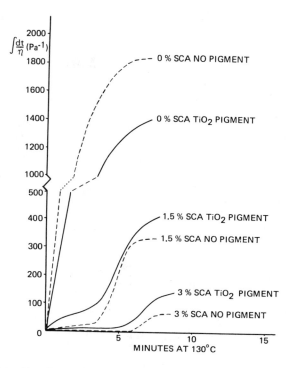

FIGURE 11.13 Sagging properties of white pigmented (p/b ratio 0.67) and unpigmented HS polyester automotive topcoats containing 0, 1.5 and 3 percent of HDI-BA on resin solids. Paints adjusted to 40" DIN cup 6.

TABLE 11.3
Solids Content of White Pigmented HS Polyester Automotive Topcoats Containing HDI-BA as SCA at 30" DIN Cup 4 and 23°C

HDI-BA on binder solids (% by weight)	Solids content paint calculated (% by weight)	Solids content paint after 1 hr at 130°C (% by weight)
0	77.6	72.7
1.5	75.2	70.7
3	72.1	67.5

217

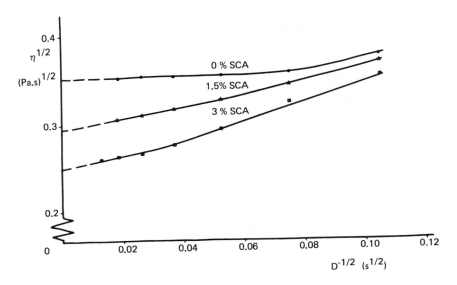

FIGURE 11.14   Casson plots of white pigmented HS polyester automotive
topcoats containing HDI-BA as SCA and adjusted to 30"
DIN cup 4 at 23°C (Table 11.3).

In general, it can be said that the decrease in solids content due to
3% by weight of SCA on binder solids was not more than 1%-2% weight
solids when the formulations were adjusted to the same viscosities
measured at a shear rate occurring in the spraying process.

STORAGE STABILITY

The storage stability in HS polyester automotive topcoats was
measured in two ways.

In Unpigmented Formulations

The SCA containing unpigmented formulations were stored in glass
tubes at 60°C and the number of days determined before the solution
visually became clear. The results are summarized in Table 11.4. These
results indicate why the HDI-CHA and HDI-StA SCAs gave paint films with
lower gloss readings when baked at 130°C (Table 11.2). Apparently they
did not disappear during the curing cycle.

TABLE 11.4

Storage Stability at 60°C of SCAs in 3% by Weight on Binder Solids in an Unpigmented HS Polyester Automotive Topcoat

Code SCA (Table 11.1)	Number of days before the unpigmented formulation became clear when kept at 60°C	
	Blocked acid catalyst	Unblocked acid catalyst
HDI – BA	~21	4
HDI – CHA	>21	>21
HDI – PrA	2	1
HDI – BuA	2	1
HDI – AmA	3	1
HDI – HeA	3	2
HDI – OcA	12	3
HDI – StA	>21	~21

In Pigmented Formulations

    Storage stability tests with pigmented formulations were only carried out with formulations containing a blocked acid catalyst. Experience with different pigmentations showed that the results with the white pigmentation were indicative for the other pigmentations. Moreover the white pigmented formulations needed the highest amount of the most efficient SCA HDI-BA. So only the results of the last combination are summarized in Table 11.5.
    For a quantitative determination of the SCA and for making SEM pictures the unpigmented formulation was included. The SEM pictures showed no decrease in particle size (particle diameter), nor a change of the particle shape after storage.
    From the work on the antisagging efficacy and the storage stability, it can be concluded that within the series of SCAs investigated, HDI-BA gave fully satisfactory results. The use of this SCA in HS automotive polyester topcoats did not have any adverse effect on the mechanical and chemical properties of the films, nor did it change the results of accelerated weathering tests and 2 years exterior exposure in Holland and Florida. After storage of the paint (Table 11.5), the film properties essentially stayed the same. By using this SCA the withdrawal of the paint from sharp edges [11] was suppressed to an acceptable level.

TABLE 11.5

Storage Stability of a White Pigmented (p/b Ratio 0.67) HS
Polyester Automotive Topcoat Containing 3% by Weight
HDI-BA on Solid Binder
Blocked Acid Catalyst

Paint properties	Temperature				
	Ambient	50°C	35°C	35°C	Ambient
	Start	4 Weeks	3 Months	6 Months	6 Months
Viscosity DIN cup 4 (S)	29.8	47.5	42.8	44.5	32.4
Solids content[a] (% by weight)	70.2	68.4	68.2	67.4	68.6
Film thickness[b] drawn film sagging test (μm)	46	53	47	55	52
HDI-BA[c] in formulation (%) (100% at start)	100	–	87.1	76.4	94.0

[a]After adjustment to 30 ± 0.5 sec DIN cup 4 at ambient temperature and
after 1 hr 130°C.

[b]Procedure on page 213.

[c]Gravimetric determination in the analogous unpigmented samples.

On the laboratory scale, HS polyester resins for automotive topcoats
could be modified with HDI-BA reproducibly without particular problems.

DISCUSSION

The concept of the Sag Control Agents proved to be very versatile for
the rheology control in high performance, high-appearance stoving automotive topcoats.

The solution of the sagging problem was reduced to designing an SCA with: (1) a particle form and size adapted to the medium for a high efficacy as antisagging agent and not hindering the levelling and flow; and (2) a reactivity for the medium that guaranteed complete physical disappearance of the particles in the curing cycle after sag control was no longer needed. The SCA should not change dramatically in amount and in particle form and size during storage.

Changing the chemistry of the SCAs was an efficient tool to succeed in the model case of an HS polyester automotive topcoat. When needed the efficacy of the SCA could be improved significantly by properly choosing the conditions of preparation. This approach offered a wide scope of possibilities by considering diisocyanates other than hexamethylene diisocyanate and a variety of amines or mixtures of amines. This array of possibilities was narrowed down by the requirements of the film properties.

The results obtained with the HS polyester topcoats supported the model for SCAs as developed in the foregoing. Especially the similarity of form and size of the collapsed structures (SEM pictures) before and after storage indicated that the nature of the particles stayed the same. Only the density of the particles decreased on storage.

The practical solutions offered by this concept of Sag Control Agents warrant further evaluation for all kinds of coating systems.

ACKNOWLEDGMENTS

We are indebted to the management of Sikkens B. V. for permission to publish this work. We express our gratitude to the coworkers of the Sikkens Central Laboratory, especially to Messrs. R. E. Boomgaard and N. Kersten for their able assistance in carrying out the experiments and to Mr. M. van Stijn for drawing the figures. We also thank Messrs. W. S. Overdiep and J. Winkeler of AKZO Corporate Research for useful discussions on rheology and for making valuable information (Sag balance results) available to us.

APPENDIX 11.1

Modification of an acrylic resin (51% by weight resin solids in xylene) with 5% by weight SCA on resins solids.

Recipe for hexamethylene diisocyanate (HDI) and benzylamine:

1 L paint tin with a diameter of 10 cm.
Cowles dissolver with a diameter of 6 cm.

| Acrylic resin | 313.73 g |
| Xylene | 75.85 g |

Add:
| Benzylamine | 4.53 g |
Cowles dissolver 1400 r.p.m.

After 5 min. add:
| 25% HDI in Xylene | 13.89 g |
Stirr 5 min at 1400 r.p.m.

408.00 g

REFERENCES

1. Zankl, E., Kreuder, H. J. and Ehring, H. Bayer AG. German patent 1669 137.
2. Beck Koller & Co., Ltd., United Kingdom patent 1230 605.
3. Marsh, F. S., Kinney, L. F. and Betty, R. J. Armour Industrial Chemical Co., German patent 1805 693.
4. Buter, R. Akzo GmbH. German patent 2751 761.
5. Dreher, J. L. and Criddle, D. W. Chevron Research Co., United States patent 3401 027.
6. August Merckens Nachf. KG. Dutch patents 7316 870, 7316 873.
7. Groenenboom, C. J., Unpublished results.
8. Buter, R. Farbe und Lack, 86:307, 1980.
9. vanDijk, J. H. VIth International Conference in Organic Coatings Science and Technology: Vol. 4 - Advances in Organic Coatings Science and Technologies Series, Technomic, 1982, p. 72.
10. Casson, N. Conference of British Society of Rheologists: Swansea, 1957; London Conference of British Society of Rheologists: Pergamon Press, 1959.
11. Discussion report on high solids coatings of the XVth Fatipec Congress Section II-4/A. Heeringa - 1980.

PROTECTIVE COATINGS BASED ON OXIDIZED BITUMEN: INFLUENCE OF EXTENDER CONCENTRATION ON RHEOLOGICAL PROPERTIES

A. Papo, V. Garzitto, and F. Sturzi

Istituto di Chimica
Università degli Studi di Udine
Udine, Italy

ABSTRACT

This work is concerned with the study of the influence of the extender concentration on the rheological properties of a series of thixotropic protective paints based on oxidized bitumen. As an extender, microtalc was used. Five paints were prepared at different talc concentrations, ranging from 0 to 32% by weight, keeping constant the thixotropic agent concentration (1.2% by weight of Bentone). Rheological tests were performed using the coaxial cylinder viscometer Rotovisko-Haake RV 11. Satisfactory results were obtained by fitting the equilibrium data of shear-stress and shear rate into an equation of the power law type. Time-dependence was investigated by applying the experimental procedure proposed by Trapeznikov and Fedotova. Additional sag resistance tests were carried out in order to confirm the validity of the rheological results. A good agreement was obtained between rheological and technological data.

INTRODUCTION

Oxidized bitumen is, at present, employed in the formulation of high-build paints, which are used as protective coatings both in the industrial and in the transportation field. These materials are designed for the

protection of metallic surfaces from corrosion caused by chemical and
stmospheric agents. Their formulation requires that also technological
aspects, such as dry-film thickness obtainable in a single coat and application properties, are to be taken into account. In fact, these are materials
which are commonly applied by means of airless spraying equipment and
allow films of high thicknesses to be laid down in a single coat on vertical
surfaces without sagging. To this end, they should have particular flow
properties. High-build paints must present low viscosity at the high shear-
rates involved during application processes, whereas high viscosity is
required at the very low shear-rates involved in film formation.

The rheological properties of high-build paints based on oxidized
bitumen have already been investigated by the authors [1-4]. In this work
the rheological investigation is extended on the study of the influence of the
extender concentration on the flow behavior in both equilibrium and non-
equilibrium conditions. Additional sag resistance tests were carried out
in order to confirm the validity of the rheological results.

EXPERIMENTAL

Materials

The materials examined were high-build thixotropic paints containing
oxidized bitumen (25-30 pen.) dispersed in white spirit (50% by weight).
As thixotropic agent, Bentone Sedapol 44, an organic derivative of mont-
morillonite, was used. As extender, microtalc (mean diameter 1.2 μm)
was employed. A series of five paints was prepared at different talc concentrations, ranging from 0% to 32% by weight, keeping constant the thixotropic agent concentration (1.2% by weight).

Paint designation and composition are shown in Table 12.1.

Apparatus and Procedure

Rheological tests. A commercial coaxial cylinder viscometer, Searle type,
Rotovisko-Haake RV 11, and a measuring device MVI (0.96 mm clearance)
were employed. Shear rates applied were within the range 2.12 - 571 s^{-1}.
Tests were carried out at 25° ± 0.1°C.

The study of the rheological properties of the paints examined was
performed by applying the experimental procedure proposed by Trapeznikov
and Fedotova [5]. This procedure consists of subjecting the material to be
tested to a given shear-rate until equilibrium conditions are attained and
then measuring the initial shear stress τ_{max} after the application of the
same shear rate after a given period of rest t_R. The procedure is repeated
for various rest times and shear-rates.

TABLE 12.1
Paint Designation and Composition

Paint	Betone wt %	Talc wt %	ϕ_B	ϕ_T
A	1.2	0	0.0058	0
B	1.2	8	0.0062	0.0256
C	1.2	16	0.0066	0.0544
D	1.2	24	0.0070	0.0872
E	1.2	32	0.0075	0.1247

With this method, it is possible to obtain both equilibrium shear stress and shear rate data and some information on the thixotropic buildup at rest of the material tested.

Technological Tests

Sag resistance tests were carried out by means of a sag index applicator. Δ_L limit thickness values correspond to the highest applicator clearance considered without sagging, curtaining or thickness variation.

RESULTS AND DISCUSSION

Rheological Tests

All the paints examined exhibited thixotropic flow behavior. The thixotropic effect was prevailing over the viscoelastic one because of the presence of both thixotropic agent and extender.

As for the equilibrium-flow data, Figure 12.1 shows the plot of shear stress, τ versus shear rate, $\dot{\gamma}$ for the paints studied. According to our earlier work [1-4], satisfactory results were achieved by fitting τ and $\dot{\gamma}$ data into an equation of the power law type (the Ostwald-de Waele model)

$$\tau = K\dot{\gamma}^n \quad n \leq 1 \tag{1}$$

where K is the consistency and n is the power index.

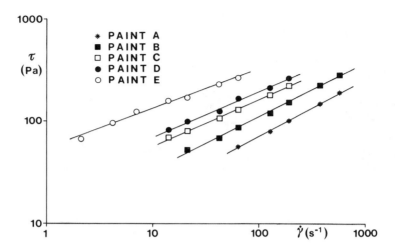

FIGURE 12.1   Equilibrium-flow curves τ versus $\dot{\gamma}$ for paints A–E.

In addition, both ln K values and shear thinning index STI (=1/n) values were plotted versus talc volume fraction Φ_T (Figures 12.2 and 12.3). Straight lines were obtained in the Φ_T range considered (ln K=1.79 +17.27Φ_T; STI=1.79+5.99Φ_T).

It must be observed that both K and STI, i.e., the deviation from the Newtonian behavior, increase with increasing talc concentration and that the addition of talc does not change the type of the rheological behavior exhibited by the system formulated without extender.

As for the time dependent behavior, experimental data obtained by applying the Trapeznikov-Fedotova procedure were treated as in our previous work [2, 3]. Since it was noticed that, at rest times higher than 15 s, the initial shear stress τ_{max} varies with rest time t_R according to an equation of the power law type

$$\tau_{max} = At_R^S \tag{2}$$

the slope S_{15} determined at t_R = 15 s was selected to characterize the thixotropic build-up at rest of the material examined.

S_{15} can be calculated by means of the following expression

$$S_{15} = \left(\frac{d\tau_{max}}{dt_R}\right)_{t_R=15} = AS\left(t_R^{S-1}\right)_{t_R=15} \tag{3}$$

Figure 12.4 shows the plot S_{15} versus $\dot{\gamma}$ for the five paints considered. It can be observed that S_{15} varies with $\dot{\gamma}$ according to equation (4).

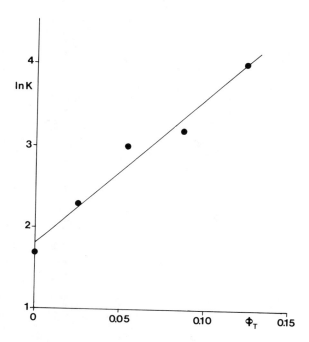

FIGURE 12.2   Natural logarithm of consistency versus talc volume
fraction.

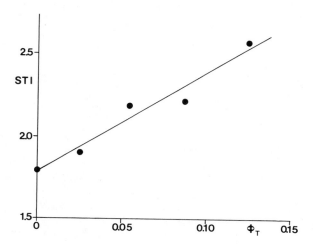

FIGURE 12.3   Shear-thinning index versus talc volume fraction.

227

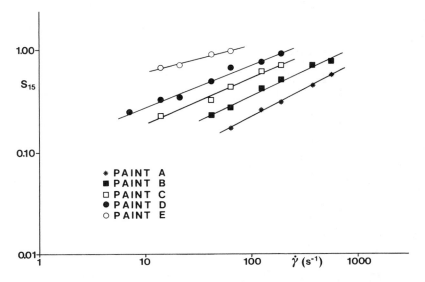

FIGURE 12.4   Build-up characteristic versus shear-rate for paints A-E.

$$S_{15} = B\dot{\gamma}^{C} \tag{4}$$

The constants B and C are given in Table 12.2.   Figure 12.5 presents the plot $(S_{15})_{\dot{\gamma}=1}$ = B versus talc volume fraction.   The results show the strong influence of talc concentration on time-dependent properties.

Technological Tests

Sag resistance results are given in Table 12.3.   As can be expected, technological tests show that the higher the talc concentration, the better is the sag resistance displayed by the paints examined.

CONCLUSIONS

A comparison between the rheological and technological tests shows that the limit thickness of sagging, Δ_L, varies with talc concentration in the same manner as the consistency K and the rebuilding characteristic S_{15}.

As already remarked [6], a reliable evaluation of the sag resistance of a paint can be obtained by coupling an equilibrium-flow parameter, which gives a measure of the viscosity level at low shear-rates, with a rebuilding

TABLE 12.2
Constants of Equation 4

Paint	B	C
A	0.019	0.537
B	0.038	0.483
C	0.066	0.447
D	0.105	0.413
E	0.333	0.257

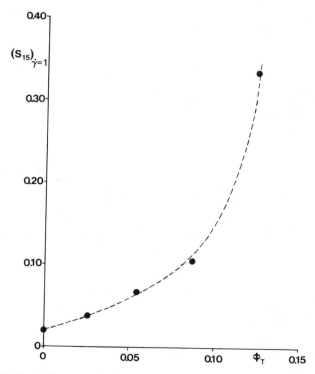

FIGURE 12.5   Influence of talc volume fraction on the build-up characteristic at $\dot{\gamma} = 1$ s^{-1}.

TABLE 12.3
Sag Resistance Results

Paint	Δ_L (μm)
A	200
B	300
C	400
D	600
E	800

one, which gives a measure of the thixotropic build-up at low $\dot{\gamma}$. To this end, both the consistency K and the build-up characteristic S_{15} proved to be an efficacious tool.

ACKNOWLEDGMENT

The authors would like to thank the SAPPI S.p.a. (Monfalcone, Italy) for supplying the materials needed for this study.

REFERENCES

1. Papo, A. and Sturzi, F. Ind. Vernice, 34(8):3, 1980.
2. Papo, A. and Sturzi, F. Ind. Vernice, 34(10):19, 1980.
3. Papo, A. and Sturzi, F. J. Oil Colour Chem. Assoc., 64(1):25, 1981.
4. Papo, A., Garzitto, V. and Sturzi, F. Paper presented at the "XV Colloque National Annuel du Groupe Francais de Rheologie," Paris, 1980. Cahiers du Groupe Francais de Rheologie, Oct. 1981, p. 149.
5. Trapeznikov, A. A. and Fedotova, V. A. Dokl. Akad. Nauk. SSSR, 95(3):595, 1954.
6. Papo, A. and Torriano, G. J. Oil Colour Chem. Assoc., 63(1):20, 1980.

A UNIVERSAL FORMULATION CONCEPT FOR PAINTS AND INKS

Palle Sørensen

Sadolin & Holmblad Ltd.
Glostrup, Denmark

INTRODUCTION

It has always been a technician's dream to have a universal formulation concept: a way to predict what raw materials should be used and the type of dispersion machinery necessary to obtain the optimum result for printing ink and paint systems. In this paper, it will be shown that it may be important to know how to select the most efficient dispersion machinery. But more important than to obtain optimal dispersing conditions will be to know the mutual interactions between pigment, binder, plasticizer, and solvent as expressed by the acid–base concept. Not only optimal dispersion conditions can be obtained by use of this concept, but also a better understanding of how to obtain optimal adhesion to a given substrate.

DISPERSION TECHNOLOGY

We all know that if we are to manufacture a printing ink we must disperse the pigment in a binder system suitable for the desired end use. But what is involved in a dispersion process? The usual answer is that you must ensure the pigment is ground to a suitable grain size so that the required color strength and gloss are obtained.

To grind a pigment in a suitable binder system, grinding machines
are used. These are of different construction, making them more or less
suitable for a given job. Some are most suitable for low-viscous systems,
some for high-viscous systems. Pigment-users are constantly calling for
better grinding equipment, and during the last few years, quite a number
of new machines have appeared that comply better with requirements for
higher yield and quality. However, with all these new developments, it is
important to make a thorough study of the dispersion technology so that
wrong investments are avoided.

Both in technical magazines [1-3], as well as from the suppliers of
the machines there is a great deal of literature, excellently describing
dispersion processes in relation to different dispersion machines. However,
most of the literature deals only with the efficiency of the machines to
physically grind the pigment to a given grain size. Even though many
authors point at the fact that physico-chemical properties like wetting are
also essential to obtaining a good dispersion, the wetting properties of the
dispersion machines have not been dealt with in detail.

When describing the mode of operation of the dispersion machinery
we normally use the term "grinding," and think in purely physical terms,
where grinding is reduction of particle size by mechanical means, usually
vigorous enough to break down sintered aggregates and even, exceptionally,
the pigment crystals themselves. Apart from this traditional way of looking at dispersion machinery, we will describe the essential importance of
the wetting necessary to obtain optimal dispersion.

Before we evaluate the relative efficiency of dispersion machinery
we have to understand the dispersion process. The dispersion process is
divided into three stages: (1) disintegration, (2) wetting, and (3) homogenization and stabilization.

1. Disintegration is the mechanical disruption of agglomerates
    into aggregates, and further of the aggregates into primary
    particles, if possible.
2. Wetting takes place by removing air and impurities, like
    water, from the pigments by means of a solvent, a binder
    or other types of dispersing agent. This interaction between
    pigment, binder, and solvent is a physical-chemical
    (thermodynamic) process in contradistinction to the grinding
    process which is essentially of a purely physical nature.
3. Homogenization and stabilization is a continuation of the
    wetting process so that sufficient binder is added in order
    to prevent flocculation after the disintegration and the
    wetting. The wetting process alone will not always give
    sufficient stabilization.

In order to describe the dispersion capacity of the machine, it is important to distinguish between the "disintegration effect" (in physical respect) and the "wetting effect" (in physical-chemical respect). Figure 13.1 shows an empirical description of the relative grinding and wetting capacity of various dispersion machines, where the viscosity of the grinding media is thought optimal for the individual machine types.

The Red Devil mixer is a good grinding machine, but it has bad wetting capacity. The ball mills give considerably better wetting, which is caused not so much by an essentially different grinding method, but is due to the fact that the ball mill is running for so long (24 h) that there is time to obtain good wetting during the grinding, whereas the Red Devil mixer normally only runs for minutes. Traditional pearl mills are better than the Red Devil mixer, despite the fact that the same grinding principle is used, because of the existence of a pressure and a controllable temperature, both of which should increase the wetting capacity of the machine.

Today pearl mills with new principles, as represented by the John mill and the horizontal mill, have been marketed. Two different principles are involved. The John mill principle can, contrary to traditional pearl mills, grind at the high viscosities used in offset formulations. The more efficient grinding is conditioned by a constant speed of the balls in the grinding zone. To achieve this the grinding zone has been moved out into the periphery of the mill (separated by a refrigeration unit) and bars have been put in, also in the outer container, giving separate grinding chambers. The construction also implies that the mill is working under a certain pressure that is determined by gravity to a smaller extent than in traditional mills. The mill operates at higher temperatures, and this together with the differences mentioned above makes the handling of high viscosity grinding bases possible.

The horizontal mill works at low to medium viscosities and runs at far greater peripheral rotor speeds than the traditional mills. Like the John mill, it is, due to its construction, working under a certain pressure. The fact that it is still working horizontally means that the gravimetrical breaking-down of the rotor is small as compared with traditional mills. The higher rotor speeds possible with this mill are comparable to that of a high speed dissolver, and result in a good combination of disintegration and wetting.

Dissolvers and Rotor Stator mixers are typical dispersion machines, which can only wet. High speed mixers can deagglomerate pigments but are normally not able to grind (deaggregate). The reason that dissolvers are said to be capable of dispersing pigment is that the surface of these pigments has been specially prepared so that it is only necessary to de-agglomerate them, since the pigments, due to their surface preparation, have the primary particle size desired. Dissolvers must also be mentioned in connection with use for predispersion. As can be seen from Figure 13.1,

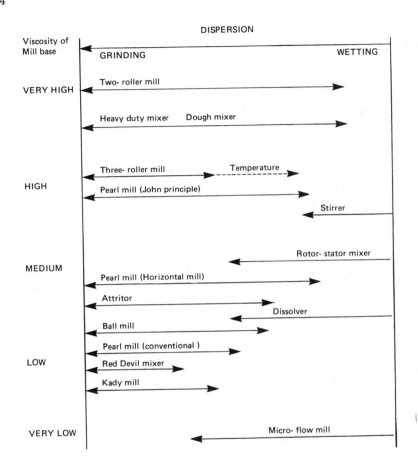

FIGURE 13.1   Relative grinding and wetting capacity of dispersing
machines.

an appropriate combination of dissolvers and grinding machines will give a
nearly ideal dispersion. It should be noted that when using dissolvers, the
dispersion should take place under clearly defined conditions with regard
to impeller size, size of vessel and peripheral speed of impeller [3].

DISPERSION EXPERIMENTS

A number of experiments have been made with different pigments
ground in three different binders using three types of dispersion machines.
One of the machines tested was a 2-roller mill, normally used for chips

manufacture, because of optimal conditions for grinding as well as wetting. The Red Devil mixer is efficient for grinding but insufficient for wetting. A ball mill may in efficiency be placed between these two.

Experiments to disperse Pigment Yellow 83 (Permanent Yellow HR) in nitrocellulose also show a remarkable difference in quality depending on the dispersing machine chosen. If the three machines are used to disperse the pigment in vinyl or polyamide, we will in polyamide observe very poor color-strength and gloss-independent of the choice of dispersion machine. In vinyl, an excellent dispersion is obtained characterized by high color-strength, gloss and transparency, but independent on the dispersion machine chosen.

These experiments show it is not possible to solve dispersion problems only by choosing the most efficient, and normally the most expensive, dispersing machine. The explanation of the results obtained will be found in the influence of the acid/base concept on formulation.

ACID/BASE CONCEPT

The interaction of pigments, polymers and solvent can be described by the acid/base concept, after Lewis, which is the interaction between electron donors and electron acceptors [4]. The dispersion results of Pigment Yellow 83 in nitrocellulose, polyamide and vinyl, in the characterization from the acid/base concept are as follows:

	Basic	Acidic	Amphotheric	Inert
Binder	Polyamide	Vinyl	Nitrocellulose	
Pigment	Diarylide Yellow PY83	Phthalo Blue PB15:4	Lake Red C PR53:1	Diarylide Yellow PY12

Best wetting will be obtained, when a basic pigment (PY 83) is dispersed in an acidic binder (like vinyl). This is the reason why the best dispersion result is obtained in vinyl regardless of the choice of dispersion machine.

If experiments as described for PY 83 are made using a sulphonated phthalo blue (PB 15:4) we find that in nitrocellulose the choice of dispersion machinery is essential. Poor results are obtained on a Red Devil mixer and the best results occur on a two-roller mill. When dispersing in vinyl poor results are obtained regardless of dispersion machinery. In polyamide excellent results are obtained with all the three dispersion machines mentioned.

The amphotheric pigment, Lake Red C, will show some response to the choice of dispersion machinery and fairly good quality to the dispersion may be found in all three binders.

For an inert pigment, a diarylide yellow, the choice of dispersion machinery is essential. Best results are obtained in nitrocellulose, and heavily flocculation is obtained in vinyl as well as polyamide.

LIMITATION IN THE USE OF THE ACID/BASE CONCEPT

The acid/base concept has in 7 years practical formulation work proved to be important in describing the mutual interaction between pigment, binder, and solvents. In a few cases the results obtained could not be explained by use of the concept. The reason was found to be the influence of polarity.

The cohesive energy density, CED, often used in the coating industry as the square root of the CED and expressed as the term solubility parameter, can be divided into three different interaction forces [5]: (1) dispersion forces, (2) hydrogen bonding forces, and (3) polar forces.

The acid/base concept which expresses the electron donor and the electron acceptor capacity (after Lewis) is mainly based on hydrogen bonding forces and may not be an expression of the total polar forces. The limitation in the use of the acid/base concept can then be observed when using binders where the polar forces are predominant. Many acrylic binders are of this kind.

CHOICE OF PLASTICIZER

In the same way as a range of pigments, binders, and solvents can be described by the acid/base forces [4] and to some extent by the polar forces, it may be essential to characterize the plasticizer by this terminology. To plasticize an acidic binder a basic plasticizer is needed, e. g., in the use of phthalate esters for polyvinylchloride. For nylon (polyamide), an acidic plasticizer is needed and often water or sulfonamides are in practical use. For acrylic binders where the polarity is predominant, phosphate plasticizer is often the best because of the much higher polarity of the phosphates compared with phthalate plasticizers.

ADHESION PROPERTIES TO SUBSTRATES

Wake [6] and Brydson [7] have stated that adhesion of coating materials to substrates can often be explained by use of the electron donor-electron acceptor properties. Again the optimal interaction occurs when an acidic binder is used for a basic substrate, and a basic binder for a substrate which may be characterized as acidic. This has also been observed by the author.

Besides the acid/base properties the polarity of the substrates is important. The hydrophilicity of the substrate, expressed by the solubility parameter or the critical surface tension, should be of the same magnitude as the coating [8, 9], Skeist [8] and later Burrell [10] also stated the importance of knowing the degree of crystallinity of plastic substrates.

CONCLUSIONS

An empirical approach for dispersion machinery which describes the relative dispersion efficiency expressed by the grinding and the wetting ability of the machines can be helpful in the selection of the right type of dispersion equipment.

To obtain optimal dispersion the interaction between pigment, binder, plasticizer and solvent, the acid/base concept is an important tool.

If the interaction parameters of the raw materials are not in agreement with the given rules, the most powerful and expensive dispersion machine cannot give a good dispersion, and only a cheap dispersion machine may be needed if the formulation is made according to the interaction parameter.

In a few cases the polarity may be predominant and in this case errors may occur if only the acid/base properties are considered.

The plasticizer should be selected in accordance with the acid/base properties or the polarity of binder used.

The adhesion of coatings to a given substrate can be explained by the acid/base concept and the polar interaction.

If the raw materials and substrates used are characterized and formulated according to the acid/base concept and in some cases also by taking the polar interaction into consideration, optimal dispersion and adhesion properties can be obtained. It can be said that we may be close to a universal formulation concept.

REFERENCES

1. Mielke, M. Farbe u. Lack, 18:495, 1975.
2. Gall, L. and Kaluza, U. Defazet, 29:102, 1975.
3. Patton, T. C. J. Paint Technol., 42:626, 1970.
4. Sørensen, P. J. Paint Technol., 47:31, 1975.
5. Hansen, C. M. J. Paint Technol., 39:505, 1967.
6. Wake, W. C. Polymer, 19:291, 1978.
7. Brydson, I. A. Plastic Materials, Ed. Newnes, Butterworths, London, 1975.
8. Skeist, I. Modern Plastics, 121, 1956.

9. Dyckerhoff, G. A., et al., Die Angewandte Macromoleculare Chemie, 21:169, 1972.
10. Burrell, H. J. Paint Technol., 40:197, 1968.

JAPANESE LAC: THE IDEAL MODEL FROM NATURAL MATERIALS

Ju Kumanotani

Institute of Industrial Science
University of Tokyo
Minatoku, Tokyo, Japan

INTRODUCTION

Japanese Lac is a naturally occurring phenolic coating material which is made from the sap of lac trees (Rhus vernicifera). The sap is stirred at room temperature, then at $20°-45°C$ for 7-8 h in a particularly-designed open vessel under exposure of an electric heater hanging over the vessel, and the endpoint of stirring is decided by judging the changing color and viscosity of the sap under treatment. The lac made usually contains 3% water and has an appropriate viscosity for coatings. The sap is used as a material for base coating, and the lac is for middle- and topcoating on wood articles, and dried in moist air for one day. The dried films are polished with charcoal, followed by coating, and drying. These same procedures are repeated several times.

In Asian countries, the same type of lacs have been made from the sap of other lac trees (Rhus succedanea and Melanorrhorea usitata). The major component is urushiol homologues, 3- or 4- substituted catechol derivatives such as laccol or thitsiol. When compared with synthetic coatings, the most prominent property of the lac is excellent durability which has been proved by use of lac wares since ancient times. In Japan, China, and some museums in the United States and Europe, there are exhibited a large number of Japanese lac properties which have been preserved over thousands of years without losing their elegant beauty.

239

Furthermore, the films in the open air are well known to degrade into powdery efflorescence from the surface of the lac films without losing beauty in appearance.

The lac is adhesive and electrically insulative. Furthermore, the lac wares are beautiful, soft in touch and yet solid. These characteristics are the specialities that are granted only to them. In terms of modern synthetic coatings, Japanese lac is a 100% solid coating belonging to an air-drying type, although the reaction mechanism is completely different from that of synthetic coatings.

This paper describes the excellent durability of lac films based on measurements of dynamic mechanical properties of the sap and lac films as a function of time, and the morphological features responsible for durability.

CONSTITUENTS OF THE SAP OF LAC TREES
(RHUS VERNICIFERA)

The sap of lac trees is a latex and water in oil-type emulsion containing 20%-25% of water, urushiol (65%-70%), plant gum (10%) and laccase (less than 1%). Urushiol, a major component of the sap is a mixture of 3-substituted pyrocatechol derivatives with a saturated or an unsaturated chain of 15 carbon atoms as shown below [1].

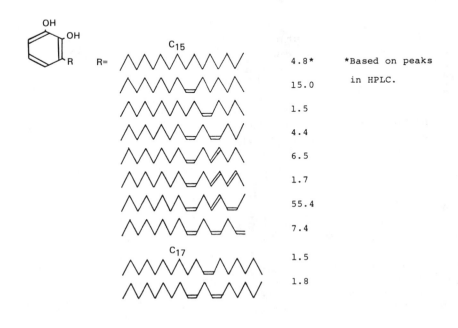

The average number of double bonds per molecule is about two for urushiol. Plant gum, the insoluble acetone part of the sap is classified into two parts: the water-soluble part (80% based on paint gum) is composed of oligomeric and polymeric sugars; the polymeric sugar exists in the salt form with a counter-cation distribution of Ca:Mg:Na = 7.8:5.5:2.8, showing a gel permeation chromatogram with two peaks corresponding to Mn 250,000 and 650,000 respectively for the dialyzed water-soluble poly-sugars [2]. Units of the acidic polymer sugars are found to consist of L-rhamnose (2% or 4%), L-arabinose (5%), 4-O-methyl-D-glucuronic acid (24%-25%), D-glucuronic acid (3%) and D-galactose (65%-67%) [3]. The water-insoluble part of the plant gum is dispersed in the oil (urushiol) phase, which is not yet elucidated. Laccase is a copper glycoprotein, p-quinol-O_2 oxidore-ductase, with Mn of 120,000 and 4 atoms of copper per molecule, consisting of common amino acids (55%) and sugars (45 mol %). Its oxidation-reduction potential and optimum pH are 415 mV (at 25°C) and pH 7.0 (hydroquinone as substrate) respectively [4].

DURABILITY OF JAPANESE LAC FILMS

The first paper with respect to durability appeared in 1961 from our laboratory. From measurements of a dynamechanical property of sap and lac films by the torsional pendulum method, it was demonstrated that the sap films caused a large extent of chemical crosslinking, whereas the lac films sustained the same degree of crosslinking during 3 years storage [5]. Recently the same conclusion was extended to lac and sap films stored over 20 years from measurements of the dynamechanical property of the films.

Figure 14.1 shows the storage modulus E' and loss modulus E" as a function of temperature at 11 Hz for the sap films J(S, 17 Y), J(B, 17Y) and the lac films J(L, 3 Y) and J(L, 17 Y) over a temperature range -150° to 200°C. It is obvious that there is a considerable difference between the sap and lac films in their dynamechanical property: the storage modulus E' (7 x 10^9 dyne/cm^2) for J(S, 17 Y) is remarkably large when compared to that for J(L, 17 Y), [(10^9 dyne/cm^2)] particularly in the high-temperature region where the modulus E' at 150°C is 7 x 10^9 dyne/cm^2 for the sap film J(S, 17 Y) and 1 x 10^9 dyne/cm^2 for the lac film J(L, 17 Y). The same relation holds true between J(S, 17 Y) and J(L, 3 Y).

Furthermore, the E" vs. temperature relation is found to be more complex for the sap film than the lac; the sap films J(S, 17 Y) and J(B, 17 Y) showed α, α', β and γ (not clear) relaxation peaks, responding to the respective E"$_{max}$ peak temperature at 120°C, 48°C and -100 to -150°C at 11 Hz, whereas, the lac film J(L, 17 Y) showed α (67°C), β (-70°C) and probably γ (below -100°C) peaks.

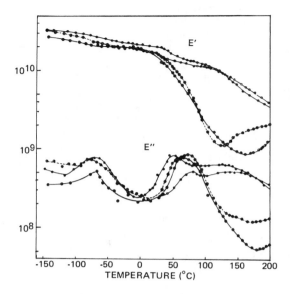

FIGURE 14.1   Storage modulus (E') and loss modulus (E") (dyne/cm^2) versus temperature (°C) at 11 Hz for Japanese lac films stored for 17 years (⊙) and 3 years (⊕), and sap film stored for 17 years (○) and baked at 110°C for 1 h and stored for 17 years (●).

From these results, it is conclusive that the sap films J(S, 17 Y) and J(B, 17 Y) are more highly crosslinked than the lac films J(L, 17 Y) and J(L, 3 Y).

The mechanical property of the sap and lac films was compared with those for synthetic-coating films. The dynamechanical profiles for the sap films are likely to be those for highly crystalline polyethylene terephthalate or highly crosslinked phenolic resins, whereas the lac films behaved like the films of phenolic resin varnishes or alkyd resins.

Assignment of these relaxation peaks was attempted from the apparent activation energy for the molecular relaxation, obtained by the Arrhenius activation plots (see Figure 14.2). The results obtained are summarized in Table 14.1.

The respective apparent activation energies obtained for the α-relaxation of J(L, 3 Y) and J(L, 17 Y), 77 and 79 kcal/mol respectively, are in coincidence with those already reported for the primary relaxation of various polymers. However, smaller varlues for J(S, 1 M) and J(B, 17 Y) (57 and 46 kcal/mol) may arise from the result of interpenetration of the α and α' relaxation peaks.

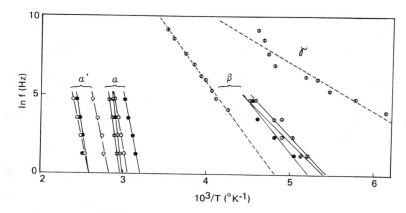

FIGURE 14.2   Arrhenius activation plots for α, α', β, and γ relaxation;
(●) J(B, 17 Y), (○) J(S, 17 Y), (◑) J(S, 5 M), (◔) J(S, 1
M), (◉) J(L, 17 Y), (◐) J(L, 3 Y), —— mechanical,
--- dielectrical.

TABLE 14.1
Dynamechanical Parameters for Sap and Lac Films[a]

	α		β		γ	
	E"max (°C)[b]	kcal/ mol	E"max (°C)	kcal/ mol	E"max (°C)	kcal/ mol
J(L, 3 Y)[b]	70	78.5	−74	10.9		
J(L, 17 Y)	68	76.5	−69	11.2		
J(B, 17 Y)	50.6	33.0	108[c]	33.0	−65	12.8

[a] 11 Hz

[b] Symbols are same as those in the text

[c] α' relaxation

As may be seen in Figure 14.2, the β-relaxation peak appears at
$-70\,^{\circ}\text{C}$ and γ-relaxation peak is likely to exist between $-100\,^{\circ}\text{C}$ and $-150\,^{\circ}\text{C}$.
The apparent activation energy for the β-relaxation peak was found to be
11 kcal/mol and is tentatively ascribed to some local-mode motions based
on the polymerized urushiol plant gum matrix involved in the sap or lac
film. The relaxation peak may be assignable to the local mode motions of
the short chains consisting of 4 carbon atoms or 3 carbon atoms and 1
oxygen atom of polymeric materials, observable in the low-temperature
region.

α AND α' RELAXATION IN THE SAP AND LAC FILMS

The origin of the α and α' relaxation peaks was clarified by measurements of dynamechanical property for the sap and lac films as a function of
storage times from 1 month to 20 years.

It is surprising that J(L, 3 Y) also shows only the α-relaxation peak,
which is almost superimposed on that of J(L, 17 Y), indicating that J(L,
17 Y) has sustained a same dynamechanical profile for about 20 years or
more without losing noticeable appearance of the α' peak.

On the contrary, the sap films behaved differently from thelac films.
In Figure 14.3 it is shown that the sap film J(S, 1 M) exhibits E''_{max} peak
at $65\,^{\circ}\text{C}$ with a shoulder at $85\,^{\circ}\text{C}$ at 11 Hz. The smaller and higher temperature α' peak is becoming larger and broader with passage of time. The
lapse of time (18 months) shifted the α' peak from $85\,^{\circ}\text{C}$ to about $118\,^{\circ}\text{C}$ in
J(S, 1 M), whereas the α' peak is still seen at $75\,^{\circ}\text{C}$. In the case of J(S,
17 Y) (see Figure 14.2), the α' peak becomes smaller and flat to a level of
the height equivalent to that of the α peak, and it's base is lying round
$200\,^{\circ}\text{C}$, but the α peak is still remaining at $75\,^{\circ}\text{C}$. In addition, the rubbery
modulus E'_{h}, a parameter of the degree of crosslinking, also becomes
larger from 1×10^{9} dyne/cm^2 to 3×10^{9} dyne/cm^2 in the lapse of 18
months.

Other data to clear up the α and α' peak are provided from the
measurements of dynamechanical property of the films of urushiol made
by baking urushiol (acetone soluble part of the sap) in air at $160\,^{\circ}\text{C}$ for 3-24
h. As may be seen in Figure 14.4, longer baking of urushiol films gives
higher modulus and higher degree of crosslinking of the resulting films. A
monotonous E'' peak which appeared at the early stage of baking at $160\,^{\circ}\text{C}$
became larger and broadened to a higher temperature region with extending
baking time; a very broad peak possessing three peaks at $75\,^{\circ}\text{C}$, $100\,^{\circ}\text{C}$, and
$150\,^{\circ}\text{C}$ is seen for the baked film for 24 h. Growing of the E'' peak in the
baking process is quite similar to those already observed in sap films as a
function of storage times as shown in Figure 14.4. These results suggest
that the appearance and growing of the α' peak in the sap films may stem
from the polymerized urushiol matrix of the sap films.

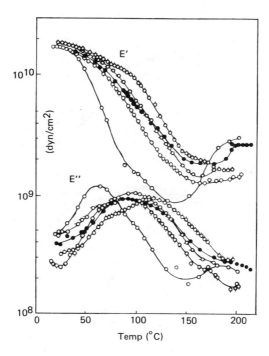

FIGURE 14.3   Variation of dynamechanical profile at 11 Hz of the sap
film as a function of time; (O) 1, (Q) 3, (●) 4, (-O) 7.6
and (◊) 13 months.

Consequently, these dynamechanical data lead to the conclusion that
the sap is inherently capable of making the α and α' relaxation peaks in
dried films, and that the smaller α' peak appears at the early stage of
drying and grows larger than the α peak as a result of oxidation followed
by crosslinking and degradation in the α' peak domain with progress of
time, whereas the α peak remains in a limited, narrow temperature
range 65-75°C.

Moreover, the lac films that show only an α peak have kept the same
dynamechanical profile for more than 20 years. The α peak temperature
in the sap and lac films indicates that the α peak is common in the both
sap and lac films. Therefore, the sap films are likely to be composed of
two domains which are responsible for the α and α' relaxations dynamechanically.

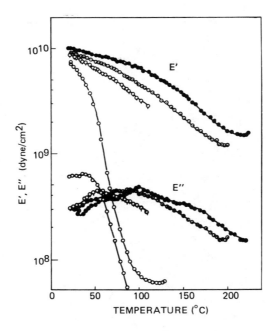

FIGURE 14.4   A dynamechanical profile at 11 Hz for a baked urushiol
film at 160°C as a function of time; (O) 3, (Q) 4.5, (⊙) 6
and (●) 24 h.

MORPHOLOGICAL FEATURES OF THE SAP AND LAC FILMS

Sap of the lac trees is dark in the polarized optical microscope, but
shows a large number of spherical crystalline grains with the maximum
size of 10 μm when dried into solid films.   This result is considered to be
reasonable.   The sap is a water-in-oil type latex, and the major part of
the gum is dissolved in the dispersing water in the oil (urushiol) phase,
which may be responsible for the formation of spherical grains.   Electron
microscopy (SEM) also shows the spherical grains in the dried-sap film
as a disperse phase.

Since urushiol is an oxidizable phenolic compound, the variation of
the dynamechanical property of the sap films is easily understandable
as a result of air-oxidation of the polymerized urushiol domain of the sap
films during storage.

On the other hand, no variation in the dynamechanical properties of
the lac films is peculiar.   SEM photographs of a section of the lac films
indicate no particular structure, however, after being ion-etched with

ionized air, the lac film is seen to contain aggregated grains, the average size being about 0.1 μm.

In order to think deeply about the particular structures of the lac films, the experimental results obtained are summarized.

1. High durability of the lac films is proved as a function of time up to 17 years dynamechanically.
2. Lac films have retained a single relaxation peak for 17 years.
3. Structure of the sap and lac films should be distinguished despite their same origin.
4. The higher barrier of polysaccarhides (plant gum) toward oxygen diffusion is well known, which is decreased by humidity absorption.
5. Urushiol is oxidizable, but should be protected from oxidation in the lac films.

In view of these results, the author proposes a cell model for the lac-film structure in which the wall of the cell consists of polysaccharides and polymerized urushiol is inside the cell. In this model, the polysaccharide wall, which may absorb humidity, keeps out humidity by the effect of hydrophobic, polymerized urushiol linked to hydrophilic polysaccharides, resulting in the polymerized urushiol in the inside of the cell being prevented from oxidative degradation.

FORMATION OF THE CELL STRUCTURES

Alternation of Viscoelastic Properties of the Sap
During the Lac-Making Process

Dynamic rigidity (G') and viscosity (η') were measured using a coaxial, double cylinder rheometer and an audio rheometer. The frequency dependence of G' and η' was obtained at an angular frequency range from 10^{-1} to 10^3 sec^{-1}.

In Figure 14.5 the measured frequency dependence of G and η' is shown for the sap. In the frequency range from 10^{-1} to 10^3 sec^{-1}, the frequency dependence of η' was found to be very large, so that the equilibrium value of η' in the very low frequency range corresponding to the zero shear viscosity could not be obtained. On the other hand, unexpectedly the frequency range was found to be smaller than that in the audio-frequency range. These phenomena suggest that a relaxation mechanism with long relaxation times may exist in the behavior of the sap. This viscoelastic behavior may result from the characteristic of the sap as an

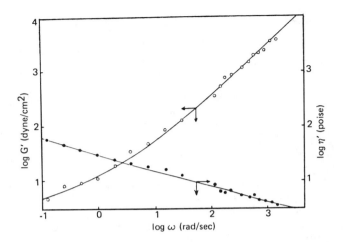

FIGURE 14.5   Frequency dependence of G' and η' for the sap of lac trees.

emulsion, namely fine droplets of an aqueous solution of water-soluble plant gum dispersed in an oil (urushiol) phase are interacting with each other, and capable of making a temporary network structure.

The effect of the lac-making process on the viscoelastic property of the sap is shown in Figures 14.6 and 14.7, the frequency dependence of G' and η' for the sap and lac being associated with oxidation in air at 40°C for 2 h (sample I), 7 h (sample II) and 7 h + heating at 90°C for 6 h (sample III). It is evident from Figure 14.6 that the values of G' at the lower frequency decrease with progress of the lac-making process. Figure 14.7 shows that the frequency dependence of η' at lower frequency also decreases with progress of the lac-making process. For sample III, as in the case of the lac, the equilibrium value of η' corresponding to zero shear viscosity is observed in the low-frequency range.

From these results, it is thought thatthe temporary network structures with long relaxation times in the sap disappeared with the progression of the lac-making process; the viscoelastic mechanism of the sap is varying as the lac-making process proceeds.

Figure 14.8 shows the frequency dependence of G' and η' for the lac made from the sap through the process. The viscoelastic behavior seems to be intermediate between the prediction for rigid and flexible chains in solution. Especially, it is emphasized that in an audiofrequency range, the values of η' seem to approach an equilibrium value higher than that of urushiol as solvent or medium.

In solutions of rigid polymer or polymer in highly viscous solvents, a nonvanishing limiting value of dynamic viscosity is observed in the high

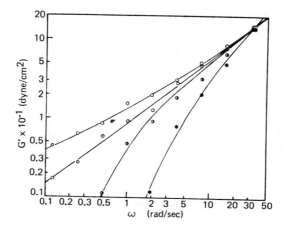

FIGURE 14.6   Frequency dependences of G' for solutions of sap (O) and oxidized for 2 h (⊙), 7 h (Φ) and 7 h + heating at 90°C for 6 h (●).

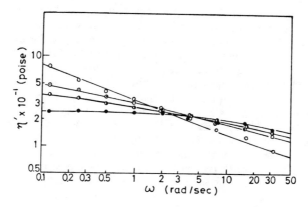

FIGURE 14.7   Frequency dependence of η' for solutions of sap (O) and the oxidized one, for 2 h (⊙), 7 h (◑) and 7 h + heating at 90°C for 6 h (●).

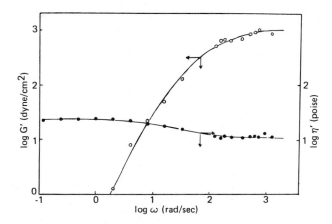

FIGURE 14.8   Frequency dependence of G' and η' for the solution of
Japanese lac.

frequency range. This phenomenon can be explained in terms of the
internal viscous force by which polymer molecules resist the rapid change
of shape.

It is thought that as the lac-making process proceeds, water molecules gradually evaporate, and the molecular chains of polysaccharides
may take a more contracted conformation. Presumably the restricted
segmental motion of the polysaccharides may contribute to the peculiar
viscoelastic behavior of the lac at higher frequencies. In other words, in
the lac-making process, water soluble plant gum which is insoluble in the
oil (urushiol) phase becomes oil-soluble, dispersing in the lac.

Plant gum, separated as an acetone-insoluble part of the sap, showed
an IR spectrum characteristic of polysaccharides in the salt form, however,
the acetone-insoluble part of the lac showed an IR spectrum superimposed
on those of the polysaccharides and urushiol.

Furthermore, the sap is partly soluble in toluene but turned into the
lac which is soluble in toluene completely via the lac-making process.
These results are in good agreement with the explanation that the plant
gum which is soluble in water and insoluble in urushiol, becomes soluble
or dispersible in the oil (urushiol) phase as a result of grafting of urushiol
to the plant gum in the lac-making process, though the bonding site
between the plant gum and urushiol molecules is not yet revealed.

Polymerization of Urushiol. Participation of urushiol in the sap- or lac-
film formation is significant, because urushiol is a major component of the
sap or lac. As shown in the following scheme,

it is established that a) urushiol (1) is oxidized into urushiol quinone (2) in the presence of laccase; b) the formed quinone (2) undergoes C-C or C-O coupling reaction with the catechol nucleus or unsaturated side chain of urushiol, giving dimeric urushiol (3, 4 and 5); c) water molecules existing in the lac-making process seem to take an important role in the determination of reaction pathway of the quinone or the reaction rate. The catechol nucleus of this dimeric urushiol is further oxidized into the corresponding quinone, followed by the subsequent coupling reaction which is shown in making dimers (3, 4 and 5). Growing of urushiol in molecular weight is reached at 20,000-30,000 (by GPC) in the lac-making process [8].

In the enzymatic oxidation of the sap or lac, molecular oxygen is trapped by the reduced laccase due to its high affinity, and consumed in a quick turnover process of laccase, and the laccase is inevitably a high potential catalyst for the oxidation of catechol to the corresponding quinone. Thus the most favorable oxidation of urushiol to urushiol quinone precedes instead of an attack of oxygen molecules at the double bonds in the unsaturated side chains.

It should be noted that the C-C coupling product (5) may make network polymer chains in which the rigid catechol nuclei are linked between flexible aliphatic chains consisting of 7 C-14 C; probably this structure is contributing to the property of the lac films with balanced rigidity and flexibility or toughness [9].

Regarding the mechanism of making the proposed cell structures from urushiol-grafted plant gum, the following two terms seem to be significant driving forces under the enzymatic film formation conditions:

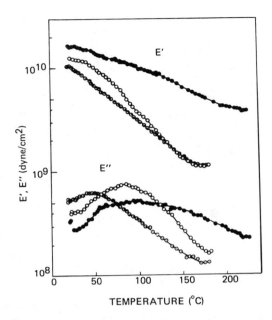

FIGURE 14.9   A dynamechanical profile at 11 Hz for a sap film stored
for 1.5 months (O), a film made by baking at 160°C for
6 h of a sap film dried over 1 week (●), and that by baking
at 160°C for 5 h of a sap directly (⊙).

TABLE 14.2
Specific Gravity of Sap and Lac Films at 25°C

J(S, 3 M)[a]	1.128
J(S, 19 Y)	1.217
J(L, 19 Y)	1.185
J(L, 1 Y)	1.180

[a]Symbols are same as in text.

(a) phase separation in the lac film due to a large difference in solubility parameter between hydrophilic polysaccharides and hydrohobic urushiol; and (b) enzymatic film formation. The former concept is understandable tentatively from the general consideration of morphology of grafted polymeric material system. Regarding the latter term, three films are compared in Figure 14.9. The difference in the physical property based on the crosslinked structures of the sap films is obviously observed between the film made enzymatically at room temperature, and that made by baking of the sap in air at 160°C for 6 h. The enzymatically-dried films showed higher storage modulus as a function of temperature than the baked film. Moreover, the lac films are found to show higher specific gravity than the others (see Table 14.2), and crystallinity can be observed in both sap and lac films, but not in the baked sap film.

REFERENCES

1. Yamauchi, Y. and Kumanotani, J. Presented at 28th IUPAC Congress, 1981: J. of Chromatgr. 243:71, 1982.
2. Kumanotani, J., Achiwa, M., Oshima, R. and Adachi, K. Cultural Property and Analytical Chemistry, p. 50, 1979.
3. Oshima, R. and Kumanotani, J. Presented at 28th IUPAC Congress, 1981.
4. Nakamura, T. Biochim. Biophys. Acta, 30:44, 1958.
5. Kuwata, T., Kumanotani, J. and Kazama, S. Bull. Chem. Soc. Japan, 34:1678, 1961.
6. Kumanotani, J. and Achiwa, M. Progr. Polym. Physics. Japan, 629, 1978.
7. Amari, T., Kumanotani, J. and Achiwa, M. Shikizai, 53(11):629, 1980.
8. Kumanotani, J., Kato, T. and Hikosaka, A. J. Polym. Sci., Part C, 23:519, 1968; Kato, T. and Kumanotani, J. Polym. Sci., Part A-1, 7:1455, 1969; Kato, T. and Kumanotani, J. Bull. Chem. Soc. Jpn., 42:2375, 1969; Kumanotani, J. and Kato, T. IUPAC International Symposium on Macromol. Chem., Budapest, 1969, vol. 5, p. 105; Kumanotani, J. Makromol. Chem., 179:47, 1978; Kumanotani, J. Fatipec, p. 360, 1976.

DEGRADATION OF ORGANIC COATINGS BY PERMEATION OF ENVIRONMENTAL LIQUIDS

Satoshi Okuda

Department of Chemical Engineering
Doshisha University
Karasuma-Imadegawa, Kamigyo-ku
Kyoto, Japan

ABSTRACT

This paper concerns the degradation mechanism of organic coatings by permeation of environmental liquids, including water. Chemical degradation of polymeric materials is explained, and then the patterns of permeation are classified and some examples are presented. Experimental results on the permeation and diffusion of environmental liquids under the influence of a temperature gradient are presented. The life of an organic coating under the influence of environmental agents is discussed. In addition a possible monitoring method for chemical degradation of the coating layer is suggested.

INTRODUCTION

Organic coating is one of the most convenient methods for protection from corrosion. The formation of a coating film is generally a complicated process, and it is very difficult to obtain uniform properties for the film. Furthermore, the coating process usually depends on the skill of trained fabricating workers. Up to this time, the durability of a coating film has been relatively short, although it is very easy to cover degradation damage by recoating or repainting.

255

But more recently, the situation has changed by the appearance of large constructions that need long-life coating, and the development of resource-saving and nonpollutive coating materials. Naturally, the durability of coating and the performance of coating films in corrosive and severe environments have been of interest in relation to the durability.

We present here results of a fundamental approach on the degradation of coatings under the influence of water and other environmental agents.

PATTERNS OF CHEMICAL DEGRADATION OF POLYMERIC MATERIALS

There are many patterns and overlapping causes of degradation of an organic coating film, but the representative patterns are blistering, delamination, decoloring or discoloration, chalking, cracking, surface solution or complete solution, and swelling. Generally, the degradation of polymeric materials can be divided into the following types: degradation by oxidation, heat, ozone, high energy radiation, high voltage electricity, microbes, mechanical energy and chemical environments. In this paper, chemical degradation is mainly considered, but other causes usually have a relationship with the chemical environmental effects.

The appearance change caused by the degradation is very different in each case. For example, the crack formed may be classified as follows.

1. Large tortoise-shell form crack. The size of crack which is about 1 cm^2, appears when the coating film is attacked by chemical reagents, and when the residual stress in the film is high and the thickness of the film is considerable.
2. Small tortoise-shell form crack. The size of crack is less than 1 cm^2, and appears when a thin coating layer is attacked by chemical regaents.
3. Straight-line crack. This appears when the coating film is thick, the residual stress is great, and the substrate body is badly deformed.

As a result of degradation (a) change of weight, (b) change of volume or shape, (c) change of mechanical properties (tensile strength, elastic modulus, hardness, etc.), and (d) change of physical properties (electric, thermal, and optical, etc.) are recognizable.

CHARACTERISTICS OF DEGRADATION BY CHEMICAL ENVIRONMENTS

The fundamental motivation of chemical degradation of a polymeric coating film is the permeation of environments (gas, vapor, and liquid) into polymeric materials. The troubles that occur when environmental agents are absorbed into polymeric materials might be classified into three main ways as shown below.

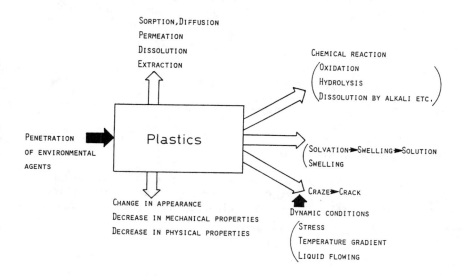

So degradation or chain breakage of polymer bonds occurs by chemical reaction with environmental agents. When such a chemical reaction is active, this material should not be used from the beginning. But, even when such a severe reaction does not exist and it is supposed to be able to use that material in some environment, other troubles often appear. One of these is solvation or swelling and the other is generation of a craze and crack. These degradations are preceded by the time effect of the environment's diffusion. When the permeation of an environmental liquid is great, molecular bonds are weakened and solvation or swelling is induced.

Next, when stresses exist, very peculiar degradation patterns can be seen. That is, the local molecular orientation induces crazing, and the craze sometimes develops to a crack. Of course, it may be regarded as degradation for a coating film when an environment penetrates through the thickness of coating layer and the function as the protective coating is lost. The dynamic conditions such as stress (not only external applied stress, but also internal inherent stress such as residual stresses are included), temperature gradient in the material, and flowing of environmental liquid are very important influencing factors for the above-described chemical degradation pattern of coating materials.

PERMEATION BEHAVIOR OF ENVIRONMENTAL LIQUIDS

The Pattern of Permeation and the Life of the Coating

There are many patterns for the sorption curve of the environmental agent into a coating layer, as shown in Figure 15.1 which are classified as follows.

Fickian diffusion
  Ideal Fickian diffusion   Curve 1
  Concentration dependent diffusion   Curve 2
Nonideal diffusion
  Time dependency of surface concentration   Curve 3
  Case II diffusion [1]   Curve 4
  Swelling diffusion   Curve 5
  Passivative diffusion

When a diffusion analysis is needed from any experimental sorption data, it is easy to consider that the diffusion process is Fickian and the coefficient of diffusion D is approximately constant. Fick's second law is expressed by equation (1).

$$\delta c/\delta t = D\delta^2 c/\delta x^2 \tag{1}$$

where c is the concentration; $\delta c/\delta x$ is the concentration gradient from the surface of specimen in the x direction perpendicular to the surface; t is time, and D is coefficient of diffusion. If q(t) is the weight change of the specimen of thickness ℓ at time t, and Q denotes a saturated weight change

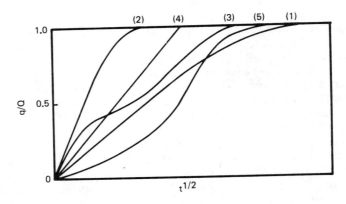

FIGURE 15.1   Classification of sorption curves (see text for detail).

after sufficient time, then the solution of equation (1) can be expressed approximately by equation (2) [2].

$$\frac{q(t)}{Q} = 1 - \frac{8}{\pi^2} \sum_{n=0}^{\infty} \frac{1}{(2n+1)^2} \cdot \exp\{- (2n+1)^2 \pi^2 Dt/\ell^2\} \tag{2}$$

The absorbed liquid diffuses to the interface with the metal substrate in the time expressed by $Dt/\ell^2 = 0.06$ as a concentration profile shown in Figure 15.2. Figure 15.3 shows the relation between a change of resistance and the generation of blisters for an epoxy resin coating (thickness = 0.09 mm) in an environment of 2 weight % hydrochloric acid. The time that the environmental liquid diffuses to the interface is $Dt/\ell^2 = 0.06$, but the time that a blister is found was $Dt/\ell^2 = 0.6$. Generally, there is a tendency that the blister appears after ten times or more than the ideal diffusion time and it depends upon the conditions of the interface with the substrate material.

There is a way to obtain the diffusion coefficient from the measurement of delay time τ in the diffusion process [3].

$$\tau = \ell^2/6D \tag{3}$$

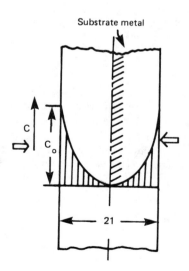

FIGURE 15.2   Concentration distribution of Fickian diffusion when an environmental liquid permeates to the center (substrate metal surface).

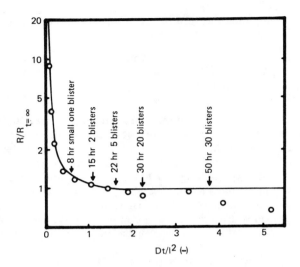

FIGURE 15.3   Relation between resistance decrease and generation of
blister (dipping test of epoxy resin coating to 2 wt % HNO_3;
thickness $1 = 0.009$ cm, diffusion coefficient $D = 6.08 \times 10^{-6}$
cm^2/h).

Menges [4] proposed the way to assume the life, L, of coating as an
application of equation 3, as follows.

$$L = \ell^2/6D + \Psi(P_i, \sigma_n) \tag{4}$$

where the first term corresponds to the time in which the environmental
liquid permeates to the interface of the substrate, and Ψ is lifetime during
which the coating layer can be kept safely after the time corresponding to
first term; Ψ is a function of inner pressure P_i which exists at the interface with the substrate; and σ_n the adhesive strength of the coating layer to
the metal substrate.

EXPERIMENTAL RESEARCH ON DURABILITY OF COATINGS

Specimen

Carbon steel plates coated with polyethylene by the fluidized bed-
coating method was used as the specimen. The size of the steel plate was
80 × 80 × 3 mm (thickness = 3 mm), and the coating thickness 0.55 ~

0.80 mm or 1.15 ~ 1.40 mm. Before testing specimens were kept in the desiccator for two weeks.

Static Dipping Test Apparatus. Figure 15.4 shows the testing device. A glass cell is used for the test of permeation from one side of the coated specimen. The clearance between the glass cell and the lining plate was electrically sealed by fluorine rubber and bound to each other by four clips. Then the environmental liquid at 40°C was poured into the cell and the device was kept in the constant temperature air bath of 40°C. The tested environments were distilled water, 10 weight % acetic acid, 10 weight % nitric acid, and 10 weight % sodium hydroxide. After a definite time, the environmental liquid was absorbed, the inside of the cell was washed with distilled water, and then the weight change of the specimen was measured.

Dipping Test Apparatus Under Temperature Gradient. The testing apparatus is shown in Figure 15.5. The specimen of polyethylene coating is fixed in the flange of the middle cell where the surface temperature of the coating is controlled at 40°C. The high temperature side is controlled at 65 ~ 67°C by mixing with a magnetic stirrer. On the other side, city water of mean temperature 10 ~ 15°C flows through the low temperature cell. The uncoated side of the metal substrate is covered with aluminum foil to prevent corrosion of base metal. The weight change of the specimen is measured.

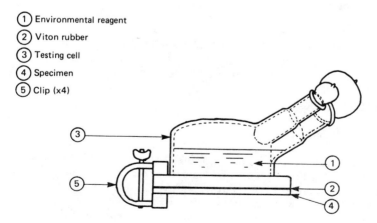

① Environmental reagent
② Viton rubber
③ Testing cell
④ Specimen
⑤ Clip (x4)

FIGURE 15.4   Experimental apparatus for the weight change test.

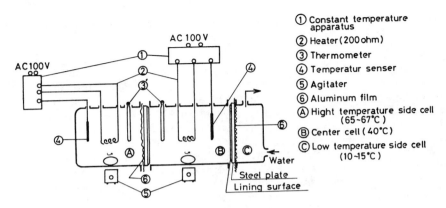

FIGURE 15.5   Diagram of immersion test under a temperature gradient.

Electric Resistance Measuring Device for the Coating Layer. A positive
platinum electrode is inserted into the mouth of cell as shown in Figure 15.6,
and the negative electrode is set on the metal substrate. The electric
resistance between the two electrodes is measured by an electrometer.
Electrical insulation was maintained between the coated plate (specimen)
and the glass cell and also the base of specimen. The cell is fixed with
four clips, and the environmental liquid is poured into the cell. The
moment the liquid was put into the cell was taken as zero time, and suc-
ceeding measurements were carried out according to a definite time interval. The reading of resistance was taken from the value at 1 min after the
indicator of resistance began to move.

EXPERIMENTAL RESULTS AND CONSIDERATIONS
ABOUT THE PERMEATION OF ENVIRONMENTAL
LIQUIDS INTO THE COATING

Static Dipping Test

The weight change when a polyethylene coated plate was dipped into
various environmental liquids is shown in Figure 15.7. The slope of the
lines are about 1/2 for the cases of distilled water and 10 weight % nitric
acid, indicating that the sorption process should be Fickian ideal diffusion,
as seen from Figure 15.1. For 10 weight % acetic acid, the slope is
almost 1.0 after 200 hours immersion. For 10 weight % sodium hydroxide,
there was almost no change in the weight of the specimen. After 3800
hours immersion in 10 weight % acetic acid, the specimen was checked for
its adhesion state to the substrate. This examination confirmed that

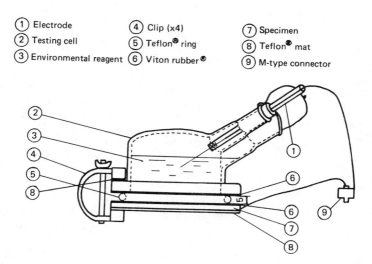

FIGURE 15.6 Experimental apparatus for the electric resistance test.

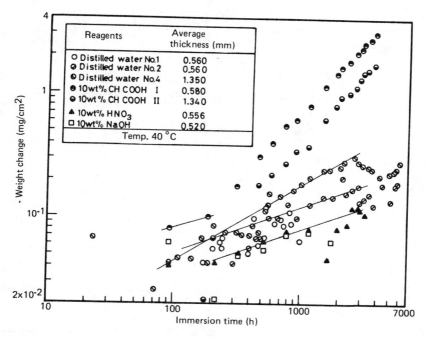

FIGURE 15.7 Sorption of environmental reagents into polyethylene lining (one-side immersion).

debonding could be detected on the surface of the metal substrate. It can
be assumed that the corrosion beginning time would be later than 200 hours,
because the slopes after 200 hours in acetic acid were changed as shown in
Figure 15.7. On the other hand, immersion for 6400 hours in water and
3700 hours in 10 weight % nitric acid showed no changes on the surface of
the substrate and any corrosion or de-adhesion could not be seen.

Liquid Permeation Test Results under a Temperature
Gradient

As described in preceding papers [5,6], the permeation velocity is
increased when the direction of heat transfer is identical to the direction
of mass diffusion. Such diffusion of distilled water into a polyethylene
coating is shown in Figure 15.8. A log-log plot of the sorption presents
a very characteristic feature for understanding the diffusion process. In
Figure 15.8, the sorption of water reaches a saturated state in about 20
hours, and this equilibrium state continues for 20 to 200 hours, after which
time the amount of sorption increased rapidly. When the coating layer was
removed, it was found that the metal surface was corroded after 3000 hours.
From this phenomenon it was considered that the rapid increase of weight
after 200 hours is due to the corrosion of metal substrate. Therefore, the
life of the coating might be considered as about 200 hours for the condition

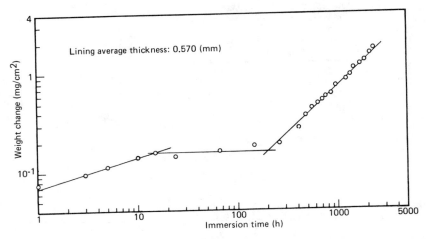

FIGURE 15.8   Sorption of distilled water into polyethylene lining under a
temperature gradient (surface: 40°C, substrate steel:
10 ~ 15°C).

that the temperature of the coating surface is 40°C and that of the substrate plate is 10 ~ 15°C respectively, and the thickness of coating is 0.57 mm. Corrosion of the metal substrate proceeds after this time.

When the direction of heat transfer and water diffusion is reversed, the relationship between sorption and time is as shown in Figure 15.9. In this case the change was not seen after 1000 hours, and any irregularity in the adhered surface was not found even after 3000 hours.

Results of Electric Resistance Measurements

Figure 15.10 shows the change of electric resistance of a polyethylene coating layer after dipping for a long period. The measured value was as high as 15 ~ 18 x 10^{10} ohm, and there was almost no change after 200 hours. After 6000 hours, the resistance value was maintained at the same value. This fact corresponds well with the results of Figure 15.7. The sorption of water is Fickian diffusion up to 6000 hours, and the coating material is performing safely in this period.

Results of Adhesive Strength of Coating Film to the Metal Substrate

Though there were no changes in electrical measurements and inspection of the corrosion state of the metal substrate, the adhesive strength of

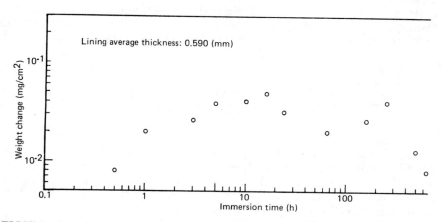

FIGURE 15.9   Sorption of distilled water into polyethylene lining under a temperature gradient (surface: 40°C; substrate steel: 65 ~ 67°C).

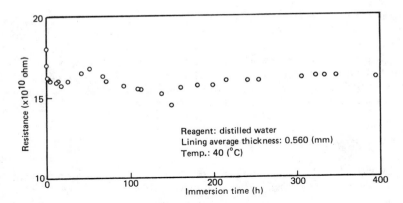

FIGURE 15.10   Relation between immersion time and electric resistance
of polyethylene lining (one side immersion).

the coating film to the substrate metal was decreased as shown in Figure
15.11.

A coating film was cut to a width of 25 mm, and the 180° piling test
was carried out by using a universal testing machine at a stretching speed
of 1.68 cm/min.   Figure 15.11 shows these experimental results.   In this
figure the first peak corresponds to the strength at the beginning of peeling.
After the first peak the strength decreases to some level but it is recovered
to its original value in the area where water permeation is not concerned.
Though we could not detect water permeation to the substrate by the measurement of weight change or electric resistance, it is interesting that the
peeling strength is already influenced by some effect of water permeation
to the interface of the substrate and is decreased in its value.

ELECTRICAL MEASUREMENT OF LINING PIPE AND
FUNDAMENTAL RESEARCH ON MONITORING
THE INTEGRITY OF LINING

Process pipes in corrosive services are frequently protected by an
organic lining. (The term lining is used instead of the term coating. Thickness of the film layer is rather large 0.5 ~ 1.0 mm.) Lining failures are
unpredictable though it is needed to predict the life of lining precisely.
Early detection is the key to any preventive maintenance program aimed at
maximizing the life of lined equipments.

Here, the electrical measuring method of coating is adopted and
fundamental research on the life of plastics linings was carried out for the
environment of 3 weight % sodium chloride (salt water), and the application
to the monitoring system of an actual polyethylene lining pipe was examined.

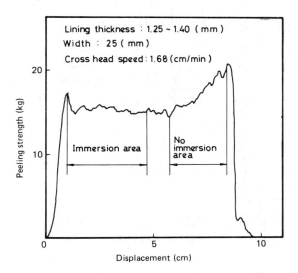

FIGURE 15.11   Relation between peeling strength and displacement (at
             21°C, after the immersion in water for 793 h).

Testing Apparatus and Testing Method

    For the measurements of their electric resistance and capacitance,
2-inch and 6-inch inner diameter polyethylene lining pipes were used. The
testing apparatus as shown in Figure 15.12 was used. The testing liquid
was 3 weight % sodium chloride and the temperature of the liquid was kept
constant at 40°C. The thickness of the lining layer initially was measured
by an electromagnetic thickness indicator, and direct electric resistance
and electric capacitance were measured by an electrometer and bridge.

Experimental Results

    Figure 15.13 shows the change of electric resistance of the 6-inch
lining pipe. From comparison of the results for 2-inch and 6-inch pipes,
it may be seen that the resistance is inversely proportional to the area of
lining. Electric capacitance also showed very little change, as shown in
Figure 15.14. Next, a testing apparatus for scale-up was designed. This
device is merely the connection of 6-inch pipe specimens as shown in
Figure 15.15. In this figure the various combinations of each pipe are
represented by alphabet letters. Accordingly, this device can measure not
only the resistance of each piece, but also the total resistance of the
connected pipes.

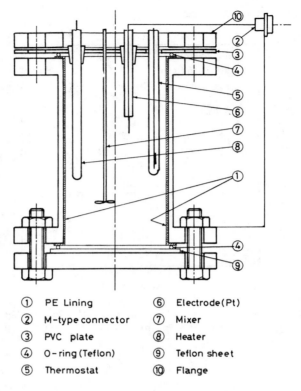

①	PE Lining	⑥	Electrode(Pt)
②	M-type connector	⑦	Mixer
③	PVC plate	⑧	Heater
④	O-ring (Teflon)	⑨	Teflon sheet
⑤	Thermostat	⑩	Flange

FIGURE 15.12   Experimental apparatus for resistance and capacitance measurements on a 6B lining pipe.

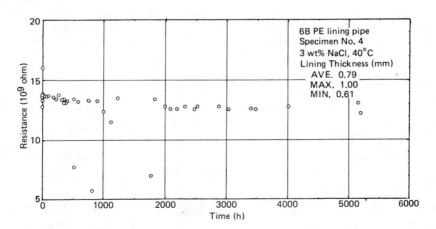

FIGURE 15.13   Relation between resistance and time.

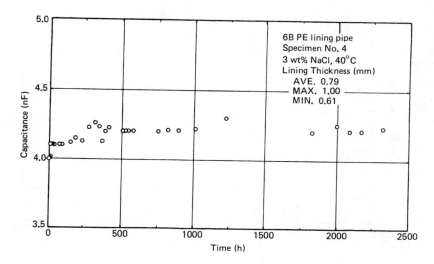

FIGURE 15.14   Relation between capacitance and time.

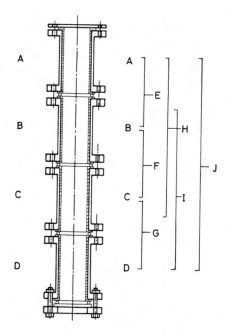

FIGURE 15.15   Combination of pipes.

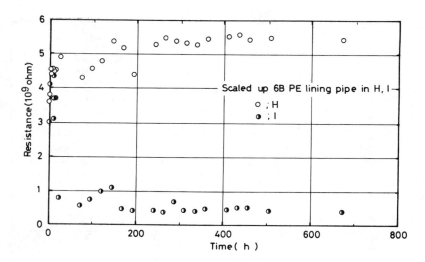

FIGURE 15.16   Relation between resistance and time of scaled up 6B
lining pipe.

The measurement in only one section A indicated very high resistance
as shown in Figure 15.13, but section D showed small resistance due to
water leakage from the flange part. If the combination contains any bad
section (Section D), it will influence the total resistance. Figure 15.16
shows an example of bad combination (I-combination) and a large decrease
in resistance. On the other hand, a good combination (H-combination)
maintained high resistance.

From these experiments it is clear that any partial defects of a big-
scale pipe will be detected by measurement of the whole system of the
equipment, and such a concept would be applicable for the monitoring
system of actual equipments.

CONCLUSIONS

From experiments on polyethylene coating and linings it was confirmed that: 1) the testing device for one side dipping test and dipping test
under a temperature gradient is very useful for the evaluation of durability
of a coating film in the presence of various environmental agents; 2) electrical measurements are also useful for the testing of durability and it can
be attached easily to the above devices; 3) from the experiments on coating specimens, the patterns and processes of the permeation of environmental liquids to the interface with the metal substrate and the blistering
mechanism have been established, and 4) it was confirmed that electrical

measurements are useful for monitoring actual equipments and their fundamental characteristics have been identified.

ACKNOWLEDGMENTS

The author wishes to express his sincere gratitude to Dr. T. Iguchi, Mr. T. Nozaka, Mr. R. Kobayashi, Mr. Y. Ito, and Mr. E. Arita for their great help with the experiments. He is also indebted to Daiichi Koshuha Ind. Co. Ltd. for cooperation in research and the manufacturing of specimens.

REFERENCES

1. Hopfenberg, H. B. and Frisch, J. H. J. Polym. Sci., Part B, Polymer Letters, 7:405, 1969.
2. Crank, J. and Park, G. S. (Eds.), "Diffusion in Polymers," Academic Press, New York, 1968.
3. Frisch, H. L. J. Phys. Chem., 63:1249, 1959.
4. Menges, G., et al., Kautschuk und Gummi-Kunststoffe, 25:213, 1972.
5. Okuda, S. and Iguchi, T. SPE NATEC (Louisville, Ken.), 120, 1975.
6. Okuda, S. and Iguchi, T. Proc. 6th Int. Conf. Organic Coatings, Science and Technology: Vol. 4 – Advances in Organic Coatings Science and Technologies Series, Technomic, 1982, p. 256.

WATER-DISPERSIBLE GRAFT COPOLYMERS
FOR CORROSION-RESISTANT COATINGS

James T. K. Woo

Glidden Coatings and Resins
Dwight P. Joyce Research Center
Strongsville, Ohio

ABSTRACT

The synthesis and characterization of an epoxy-acrylic graft copolymer is described. The grafting of acrylic monomers onto epoxy resin takes place in the presence of a free-radical initiator. The grafting is believed to be of a "grafting from" process. The graft copolymer when neutralized with base forms an excellent and stable dispersion in water. The stability of the dispersion is due to the absence of ester linkages which are susceptible to hydrolysis in water. Characterization of the graft copolymer by solvent extraction indicated that the graft copolymer consists of the following: 1) 47% of the epoxy resin is ungrafted; 2) 61% of the acrylic monomer polymerizes to form acrylic copolymer; and 3) 39% of the acrylic monomer is grafted onto 53% of the epoxy resin. From ^{13}C nuclear magnetic resonance spectroscopy, grafting appears to take place at the aliphatic backbone carbon atoms of the epoxy resin.

INTRODUCTION

Graft and block copolymers have been known for many years for their unique properties which cannot be obtained from the homopolymer alone. This paper describes our work on a graft copolymer for water-reducible systems.

Graft and block copolymers, because of the imcompatible nature of
the constituent blocks or graft, form micelles in solution. The micelliza-
tion behavior is very dependent on the chemical nature and sizes of the
blocks or graft; concentration, molecular weight, and composition of the
copolymer; the nature of the solvent; and the temperature. What we have
done is to carry this one step further, dispersing it in water. We have a
high molecular-weight epoxy resin ($\sim 10,000$), and we graft monomers onto
the epoxy resin and then the graft copolymer is dispersed in water. Grafting is done by free-radical means [1].

In free-radical grafting [2] there are two possible processes taking
place: a) Grafting from the free radical (or other active site) is generated
on the backbone and subsequently initiates the polymerization of monomers
to produce branches; b) Grafting onto a growing free radical (or other
active species) attacks another preformed polymer preferentially carrying
suitable substituents and thereby produces a branch of the preformed
backbone.

With a few exceptions [3], most graft copolymers from free-radical
induced grafting processes usually not only lead to the desired graft copolymers but also to homopolymers and other side reactions. Consequently,
the exploration and detailed characterization of grafts produced by free-
radical methods is often cumbersome or sometimes impossible even by
present-day analytical techniques.

EXPERIMENTAL

Example I: Preparation of Graft Epoxy-Acrylic
Copolymer by Carbon-Carbon Bond Formation

Commercial grade liquid epoxy resin DER-333 (Dow Chemical Company, Midland, Michigan) and epoxy-resin grade bisphenol A were used.
Monomers and solvents were of commercial grade and not purified.

A high molecular-weight epoxy resin was prepared from reacting
995 g of DER-333 with 536 g of bisphenol A in the presence of 310 g of 2-
butoxy-ethanol-1. The molecular weight of the epoxy resin was calculated
to be about 8000 from the epoxide value. 281 g of 2-butoxy-ethanol-1 and
888 g of n-butanol were then added. This mixture was heated at 116°C and
a monomer mixture consisting of 283 g of methacrylic acid, 87 g of
styrene, 285 g methyl methacrylate, and 30 g of benzoyl peroxide was
added dropwise through an addition funnel. The acid number on solids
was 85.

The grafted copolymer mixture was then fed into an agitated reducing
vessel containing water and dimethyl ethanolamine. The dispersion was
formed easily. The temperature of the resulting dispersion was $\sim 70°C$,
and the dispersion was agitated for ~ 1 h. Some more deionized water was

added so that the dispersion had the following properties: N. V. : 20%; pH: 7. 8; and viscosity: (Ford #4 cup) 22 sec.

Example II: Preparation of Graft Epoxy-Acrylic Copolymer in the Absence of Epoxy Functionalities

An agitated, nitrogen-purged reaction vessel was charged with 1, 079 g of the liquid epoxy resin DER-333, 310 g of ethylene glycol monobutyl ether, and 676 g of bisphenol A. The contents were heated to 140°C, and the heat turned off. The temperature rose to 170°C, at which temperature the reaction was held for 5 h. At the end of this time, the oxirane content was 0. 074%. 701 g of n-butanol were added and the contents allowed to cool overnight.

The bisphenol A terminated epoxy resin was heated to 117°C and then reacted with a monomer mixture that was added slowly to the terminated epoxy resin over a 2-h period. The monomer mixture was made up of 365 g of methacrylic acid, 191 g of styrene, 48 g of wet benzoyl peroxide (78% BP in water), and 157 g of ethylene glycol monobutyl ether. After this period the Acid Number of the Product was 104.

A neutralizing solution of 4248 g of deionized water, 132 g of di-methylanolamine, and 120 g of ethylene glycol monobutyl ether was used. At that point the nonvolatile content of the emulsion was 27. 1% and the viscosity as measured by a No. 4 Ford Cup at 25°C was 105 sec.

Example III: Synthesis of Epoxy-Styrene Graft Copolymer

Into a four-neck, 5-L round-bottom flask was charged 1141 g of liquid epoxy resin DER-333 of Dow Chemical Company, 614 g of bisphenol A and 310 g of 2-butoxy-ethanol-1. Through the four necks of the flask were placed the following: thermometer, nitrogen inlet, mechanical stirrer, water-cooled condenser, and a dropping funnel. The reaction mixture was heated to ~ 150°C under a 20-inch vacuum and 20 c. c. /min. N_2 purge in order to remove ~ 24 g of volatile material (consisting mostly of water and xylene from DER-333).

At 150°C, heating was stopped and an exotherm was observed. The temperature increased to 176°C and was held there for ~ 2 h. The Gardner-Holt viscosity of the advanced epoxy resin was Z_2 (40% N. V. in 2-butoxy-ethanol-1). At that time, 170 g of 2-butoxy-ethanol-1 was added to the reaction mixture followed by 826 g of n-butanol. The temperature of the reaction mixture was stabilized at 115°C, and a monomer solution of 431 g of styrene, 38. 5 g of wet benzoyl peroxide (78% in H_2O), and 62 g of n-butanol was slowly added to the epoxy resin. The addition of monomer took about 2 h, and the reaction mixture was held at 115°C for three more hours. At the end of this grafting stage, a sample was taken for non-volatile (N. V.) and acid number (A. N.) determinations. The N. V. was 55. 5% and A. N. was 5 on N. V. The A. N. was attributed to the benzoic acid formed from the decomposition of benzoyl peroxide. If all the benzoyl peroxide decomposed to form benzoic acid, the A. N. should be 6. 5. Therefore 77% of the benzoyl peroxide decomposed followed by hydrogen abstraction to form benzoic acid.

FRACTIONAL PRECIPITATION

A physical blend of 80% epoxy resin (1535. 5 g, 57. 37% N. V. [a]) and 20% acrylic copolymer (428 g, 51. 4% [b] N. V.) of methacrylic acidstyrene; 2:1 mol ratio was made. A solution of 100 g of physical blend and 150 g of NMP was made (22. 4% N. V. solution). To the first was added 155 g of toluene (which should precipitate the acrylic), to the second was added 25 g of mineral spirits (which also should precipitate the acrylic), and to the third was added 25 g of 2-butoxy-ethanol-1 and 25 g of D. I. water (which should precipitate the epoxy). The results are listed in Table 16. 1. Therefore, in this experiment with toluene, quite a bit of the acrylic copolymer ~ 50% is found in the toluene-rich, upper layer. Based on the acid number of 396. 4, the bottom layer gave a fairly good precipitation of the acrylic copolymer. (Theoretical A. N. of the pure acrylic copolymer is 424.) The experiment using mineral spirit in place of toluene gave no separation of the two components at all. The third experiment using 2-butoxy-ethanol-1 and water gave a good precipitation of epoxy resin, as evidenced by A. N. = 0 in the bottom layer. The 11. 12 g corresponds to ~ 83% recovery of the epoxy resin.

To optimize the toluene fractionation, increasing amounts of toluene were added. A solution of two parts of the blend to three parts NMP was prepared. To 100 g of this solution was added 150, 207, 300, and 500 g of toluene. The results are given in Table 16. 2 and plotted in Figure 16. 1.

It is evident from the data that a complete separation of free-epoxy resin did not occur. Very large amounts of toluene have to be used for the A. N. of the upper layer to be 0, that is, a complete separation of the epoxy resin from the blend.

For the blend, when 500 g of toluene is used the acrylic is soluble to the extent needed to give an acid number of 23. 8.

In the grafted copolymer, the presence of the graft, especially epoxy resin grafted with a small amount of methacrylic acid would raise the acid

---

[a] In 2-butoxy-ethanol-1

[b] In n-butanol

TABLE 16.1
Precipitation Results with Various Solvents

	I	II	III
Resin (blend) (56% N. V.)	30	30	30
N-methyl pyrol (NMP)	45	45	45
Toluene	155	--	--
Mineral spirits	--	25	--
2-Butoxy-ethanol-1	--	--	25
D. I. water	--	--	25
Weight of top layer	220	25	94
Solids in top layer	14. 21[a]	0	5. 23
A. N. (NVM) of top layer	41. 6	--	276
Weight of bottom layer	4. 2	75	29. 5
Solids in bottom layer	1. 55	18	11. 12
A. N. (NVM) of bottom layer	396. 4	85	0

[a] If there is complete separation of the two in the blend, i.e., the epoxy resin is formed only in the top layer (the toluene rich layer) and the acrylic copolymer in the bottom layer, there should be 13.44 g of epoxy in the top layer and 3.36 g of acrylic copolymer in the bottom layer.

TABLE 16.2
Precipitation with Increasing Toluene Levels

Wt. toluene	Wt. upper	Solids in upper	AN (NV) of upper
150	240	20. 59	50. 46
207	302	20. 29	37. 05
300	390	18. 68	30. 70
500	590	18. 88	23. 80

* Each 100 g of the epoxy acrylic blend solution contains 22.4 g solid determined by nonvolatile analysis. 17.92 g of the 22.4 g is epoxy resin and 4.40 g is acrylic copolymer.

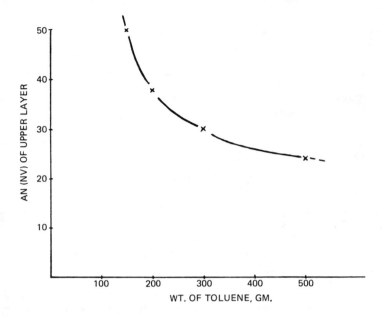

FIGURE 16.1   Fractional precipitation of a physical blend of epoxy resin,
acrylic copolymer. Weight of toluene added versus the AN
(NV) of the upper layer.

number of the upper layer. In other words, epoxy resin grafted with a
small amount of acid would behave like epoxy resin.

In the next series of experiments, different amounts of the blends
were replaced by the graft copolymer mixture and the increase of acid
content in the upper layer of toluene was recorded. The data are given in
Table 16.3. As expected, the acid content of the upper toluene layer did
increase with more replacement of the blend with the grafted copolymer
mixture. The weight of the acrylic in the upper is found by multiplying
the N. V. by 590 to give the weight of solids and then multiplying this by
the acid number on N. V. and dividing by 424. As an example for the 50/50
trial

$$\text{Wt. Acrylic in Upper Toluene Layer} = \frac{590 \times 0.0312 \times 27.83}{424} = 1.29$$

The weight of acrylic from the graft polymer is found by subtracting
the acrylic which would be expected to be in the upper layer from the blend
based on the 100% blend trial. There should be 0.98 g contribution from
the blend; in our 50/50 trial, we would expect 0.49 g to be contributed

TABLE 16.3

Blend/ graft copolymer	% NV upper	AN (NV) upper	Weight acrylic in upper	Acrylic from resin	Acrylic adj. to 100%	Increased acrylic solubility
100/0	2.82	24.89	.98	--	--	--
80/20	2.73	27.07	1.03	.246	1.23	.25
50/50	3.12	27.83	1.29	.800	1.60	.62
20/80	2.94	35.05	1.43	1.234	1.54	.56
0/100	2.89	38.8	1.56	1.560	1.56	.58

from the blend and the rest from the grafted copolymer. In the 50/50 trial, the actual acrylic is 1.29 g -0.49 g = 0.800 g. This value is then adjusted to 100% graft by dividing by 0.5 to give 1.60 g. But we would expect 0.98 g to be soluble from the blend so the increased solubility which may be due to grafting is 0.62 g. The total weight of acrylic is 4.48 g. Initially, we can calculate the percent increased acrylic solubility (PIAS) to be

$$\frac{0.62}{4.48} \times 100\% = 13.8\%$$

This PIAS is a figure that can represent the amount of graft epoxy acrylic copolymer present in a mixture where epoxy resin, acrylic copolymer and graft epoxy-acrylic copolymer are present.

RESULTS AND DISCUSSION

Molecular Weight Data

Gel-permeation chromatograms were obtained for the epoxy resin, acrylic copolymer of styrene-methacrylic acid prepared in the absence of epoxy resin, a blend of epoxy resin and styrene-methacrylic acid copolymer and the graft copolymer (Figures 16.2-16.5). As can be seen, the molecular weight of the graft copolymer is a little higher than that of the epoxy resin, and this is confirmed from computer printout, shown in Tables 16.4 and 16.5. This data will be important when the structure of the grafted copolymer is discussed. The molecular weight of the graft copolymer with different levels of free radical initiator is shown in Table 16.6. Except for

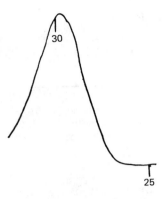

FIGURE 16.2   Gel permeation chromatogram of bisphenol A epoxy resin.

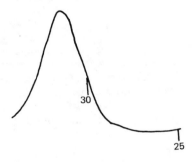

FIGURE 16.3   Gel permeation chromatogram of styrene-methacrylic acid copolymer.

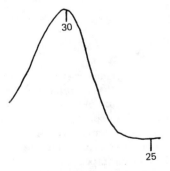

FIGURE 16.4   Gel permeation chromatogram of epoxy resin/styrene-methacrylic acid copolymer blend.

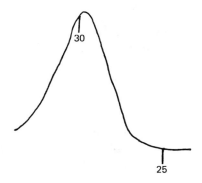

FIGURE 16.5   Gel permeation chromatogram of epoxy-acrylic graft
copolymer.

TABLE 16.4
GPC Data of Starting Epoxy Resin Component
Molecular Weight Distribution

	Number	Weight	Z
Mean	.161E 04	.805E 04	.252E 05
Variance	.104E 08	.138E 09	.759E 09
Skewness	.814E 01	.455E 01	.282E 01
Kurtosis	.147E 03	.375E 02	.128E 02

TABLE 16.5
GPC Data of Graft Epoxy-Acrylic Copolymer
Molecular Weight Distribution

	Number	Weight	Z
Mean	.186E 04	.801E 04	.309E 05
Variance	.114E 08	.183E 09	.186E 10
Skewness	.101E 02	.710E 01	.405E 01
Kurtosis	.293E 03	.104E 03	.230E 02

TABLE 16.6
GPC Data of Grafted Copolymers Made with Different
Levels of Free Radical Initiator

	% Free radical Initiator	Number Ave. Mol. Wt.	Wt. Ave. Mol. Wt.	Mol. Wt.
(1)	1	8,480	26,680	64,800
(2)	2	4,440	12,960	32,120
(3)	3	4,000	12,880	35,080
(4)	5	4,200	13,240	35,160
(5)	7	3,904	13,000	40,400
(6)	10	4,120	14,120	42,400
(7)	15	4,920	17,720	72,400

the 1% level, the molecular weight tends to increase especially in the case of weight average molecular weight and Z average molecular weight. The increase in the molecular weight from higher level of initiator is probably due to increased chain coupling.

Glass Transition Temperature Data

Glass transition temperatures of various components are listed in Table 16.7 together with the blend and graft copolymer. The significant drop in Tg of the graft copolymer seems to indicate that the grafted chains are quite efficient in increasing the degree of mobility of the epoxy polymer backbone.

When the glass transition temperatures of the graft copolymers prepared with different amounts of free-radical initiator were obtained, the data showed a distinct narrowing trend in the glass transition temperature range. This seems to indicate that with higher amounts of free-radical initiator, the graft copolymer is becoming more uniform or it may indicate that there may be a higher level of grafting present (Figure 16.6).

Particle Size Data

The particle size of the aqueous dispersion made with increasing levels of free-radical initiator show a significant decrease. The particle size data are summarized in Table 16.8.

TABLE 16.7
Glass Transition Temperature Data from DSC

	T_g
Epoxy Resin	80–85°C
Styrene–MAA copolymer	110°C
Blend	75–80°C
Graft copolymer	50–65°C

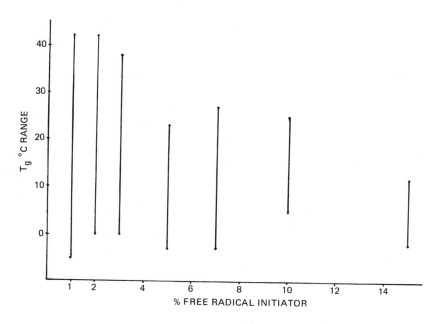

FIGURE 16.6 T_g range versus free radical initiator concentration.

TABLE 16.8
Particle Size Distribution of Dispersion from Grafted
Copolymers Made with Different Levels of
Free-Radical Initiator Concentration

	% Free radical initiator concentration	Surface diameter	Volume diameter	Specific surface	Wt. ave. diameter
1)	1	2.5675	2.7826	18000	2.1953
2)	2	1.6910	1.8393	27030	2.6304
3)	3	0.8716	0.9537	51530	2.8190
4)	5	0.5529	0.5761	94060	1.6759
5)	15	0.3108	0.3158	180400	1.1457

Aqueous dispersions of epoxy-acrylic graft copolymer made with different levels of free-radical initiator are listed in Table 16.9. There does not seem to be any correlation between the free-radical initiator concentration with the viscosity of the aqueous dispersion.

SEPARATION OF GRAFT COPOLYMER

This acrylic composition of ~2 parts of methacrylic acid and ~1 part of styrene polymerizes azeotropically. Except towards the end of the polymerization, there is a nearly random distribution of the two monomers throughout the acrylic polymer chain. This means that there is very little homopolymer of methacrylic acid or styrene. In the graft copolymer mixture after separation, a polymer having an acid value on solids of 424 would be the free-acrylic copolymer, a polymer of acid value zero is pure epoxy and a polymer of acid value in between can be assumed to be a graft copolymer of epoxy acrylic. In order to separate any free acrylic or free epoxy, it is necessary to find solvents which will act as a solvent for one component while being a nonsolvent for another. The acrylic is very polar while the epoxy is also polar but to a lesser degree. Any graft copolymer would be intermediate in polarity. Most solvents which dissolve one component will also dissolve the other two.

There are some solvents which were selective. Ketones (methyl ethyl ketone and methyl isobutyl ketone) dissolve only the epoxy; toluene

TABLE 16.9

Aqueous Dispersion of Epoxy Acrylic Graft Copolymer
with Different Levels of Free Radical Initiator

	% Free radical Initiator	Ford #4 cup viscosity (sec)	N. V.	A. N.
1)	1	47	22.8	85.9
2)	2	60	22.5	85.7
3)	3	69	23.1	90.5
4)	5	32	22.4	88.1
5)	7	62	23.0	85.6
6)	10	26	21.5	90.6
7)	15	26	22.6	93.0

and chloroform also dissolve only epoxy. The acrylic copolymer is soluble in n-butanol, water/N-methyl pyrolidone (NMP) mixtures. NMP, alcohols, tetrahydrofuran, and dimethylformamide all tend to dissolve both epoxy and acrylic. It would seem that the graft copolymer should exhibit a solubility dependent on the relative amounts of acrylic grafted onto the epoxy. A molecule with large amounts of acrylic grafted onto the epoxy would be expected to resemble the acrylic in solubility. A molecule with a small amount of acrylic grafted onto the epoxy will resemble the epoxy in solubility. For a start, a physical blend of epoxy resin and acrylic copolymer was made. The composition of this blend is same as the graft copolymer except the blend contains no graft components. The acrylic copolymer was prepared similarly as in the graft copolymer expect without the presence of the epoxy resin. Several solvent extraction techniques were used.

Fractional Precipitation

In this experiment, the idea was to dissolve the graft epoxy-acrylic copolymer in a solvent that would dissolve all three components, i.e., free-epoxy resin, free-acrylic copolymer, and grafted epoxy-acrylic copolymer. A nonsolvent for one of the components was added to selectively precipitate one component.

This technique did not separate the various components but was important in showing that the polymer is not totally grafted. It also provided data showing that the amount of grafting is dependent on the initiator level (Figure 16.7).

A physical blend of 80 parts high molecular-weight epoxy resin and 20 parts acrylic copolymer of styrene-methacrylic acid was prepared. This mixture was dissolved in a common solvent, N-methyl pyrrolidone. Using toluene as a precipitation solvent (for the acrylic copolymer), based on acid value (Table 16.1), the fractionation gave a fairly good separation of the acrylic copolymer. (Theoretical A.N. of the acrylic copolymer is 424, A.N. of the bottom layer using toluene as solvent was 396.4.) Other solvents such as mineral spirit and 2-butoxy-1-ethanol-H₂O mixture did not work well.

To optimize the toluene fractionation, increasing amounts of toluene were added, as can be seen from Table 16.2 and Figure 16.1. With increasing amounts of toluene, and A.N. of the upper layer decreases, but never approaches zero, where a complete separation of the epoxy resin

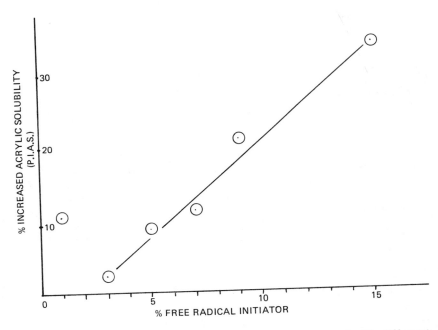

FIGURE 16.7   Solubility data of graft copolymers prepared with different levels of free radical initiator.

would have occurred. Perhaps, with an experiment where repeated extractions with toluene were used, a better separation of the epoxy resin from a blend of epoxy resin and styrene-methacrylic copolymer would have been obtained.

In the next series of experiments, different amounts of the blends were replaced by the graft copolymer, and the increase of acid contnt in the upper layer of toluene was recorded (Table 16.3). This increased acid value (called increased acrylic solubility) is due to the presence of epoxy resin grafted with a very small amount of styrene-methacrylic acid. The solubility of this type of graft copolymer would resemble that of the epoxy resin. This increased acrylic solubility was then calculated to be percent increased acrylic solubility PIAS. This technique was used on the graft copolymer samples prepared with different levels of free-radical initiator concentration.

The PIAS is plotted versus free-radical level in Figure 16.7. The values are seen to lie on a fairly straight line with the exception of the 1% free-radical initiator which had a low-monomer conversion. This linear increase in PIAS with an increase in free-radical initiator level may represent the relative amount of grafting in this epoxy-acrylic graft copolymer system.

Solvent Extractions

In order to obtain a pure sample of "free" acrylic copolymer or "Free" epoxy resin, a solvent extraction technique was used. In this procedure, attempts were made to find a solvent pair which were immiscible, and each solvent had to be a good solvent for one component and a nonsolvent for the other.

It was found that a mixture of 40 g of the epoxy-acrylic graft copolymer mixture, 160 g NMP, 160 g toluene, and 60 g of deionized water separated into 3 clear, distinct layers. The analysis of each layer is given in Table 16.10. Initially there was 4.48 g of acrylic of which 0.93 g is free; therefore, at least 20.7% of the acrylic is ungrafted. Using butyl acetate/NMP/water and MIBK/NMP/water mixtures, 1.48 g of free acrylic out of 4.48 g total for a 33% level of free acrylic was obtained.

Optimizing the MIBK extraction gave a value of 60.7% of the acrylic being free. Figure 16.8 is the GPC curve of the acrylic copolymer used to prepare the blend. Figure 16.9 is the free-acrylic copolymer extracted from the grafted mixture. It is seen that the acrylic extracted from the grafted mixture is of higher molecular weight. This could be due to a viscosity effect, such as the Trommsdorf effect, in making the grafted copolymer or the procedure may be more selective towards removing high molecular weight acrylic.

TABLE 16.10
Solvent Extraction Results with Toluene/NMP/Water

Layer	Wt. of acrylic	Wt. of epoxy
Upper (toluene rich)	1.66	18.08
Middle	2.05	1.63
Lower (water rich)	0.93	0.0

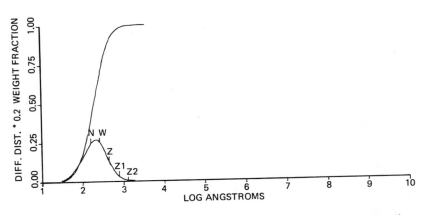

FIGURE 16.8   GPC curve of the acrylic copolymer prepared under the same condition as in grafting but in the absence of epoxy resin.

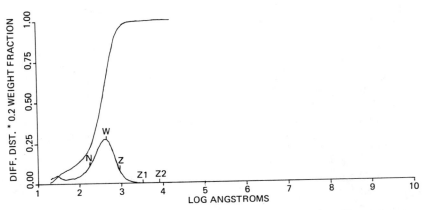

FIGURE 16.9   GPC curve of the acrylic copolymer isolated by solvent extraction.

Soxhlet Extraction

The Soxhlet extraction technique has been used to isolate [4] epoxy from poly (methylmethacrylate-g-epoxy) using petroleum ether as the solvent. An attempt was made to isolate free epoxy resin from the epoxy-acrylic graft copolymer mixture.

300 g of the graft copolymer mixture in 2-butoxy-ethanol-1 and n-butanol, ~60% N. V. was dissolved in 400 g of THF and precipitated in n-heptane. The solids were redissolved in 400 g of THF and reprecipitated in n-heptane. The isolated solids were then placed in a vacuum oven at 40°C overnight at 28" of mercury. The final resin was a white solid. It was pulverized in a Waring blender before extraction.

Using petroleum ether, no epoxy was removed after refluxing overnight. Chloroform was too good a solvent and did not give a pure sample of epoxy.

The use of toluene extracted 3.78 g of solids with an acid number of only 6 on solids after three days of extraction. This is essentially pure epoxy. This weight is 47.2% of the epoxy initially present. The infrared spectrum of the isolated resin showed no presence of carbonyl. The spectrum was almost identical to a reference spectrum of DER 669 epoxy resin (Figures 16.10 and 16.11).

Figure 16.12 is a GPC curve of the epoxy used before the acrylic grafting. Figure 16.13 is a GPC of the extracted epoxy. It is evident that the molecular weight of the extracted epoxy is lower than the molecular weight of the initial epoxy. This may indicate one of two possibilities: 1) the higher molecular weight epoxy has a greater chance of being grafted, or 2) our solvent (toluene) is a better solvent for the low molecular weight epoxy.

In order to see if the toluene is selective towards the low molecular weight epoxy a Soxhlet extraction of DER 669[R] using toluene was carried out. DER 669 has approximately the same molecular weight as the epoxy used in preparing the graft copolymer. After three days of extraction, all of the epoxy had been removed. This means that toluene is not selective towards low molecular-weight epoxy. This indicates that the higher molecular-weight epoxy has a greater chance of becoming grafted.

This is in contrast with grafting on polystyrene [5] where in grafting monomers on polystyrene with free-radical catalyst (ditert butyl peroxy oxalate) it was found by Electron Spin Resonance (ESR) that the polystyrene was selectively attached at the tertiary position. The relative reactivity of polystyrene was found to increase with decreasing molecular weight and with dilution of the solution with benzene. These results suggest that the reactivity of polystyrene depends mainly on the nature of its coiled conformation in solution. In the epoxy resin case, the molecular weight of the epoxy resin is low compared to that of polystyrene. The higher molecular weight epoxy resin tends to have greater reactivity towards grafting. One

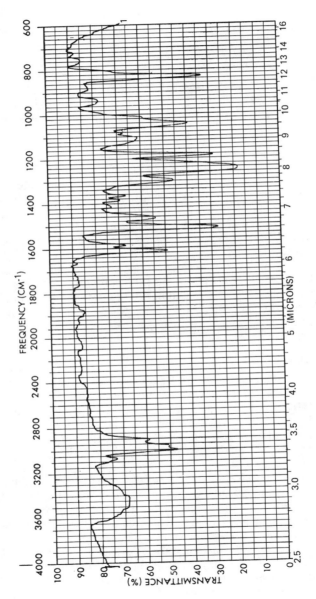

FIGURE 16.10   IR curve of extracted epoxy resin.

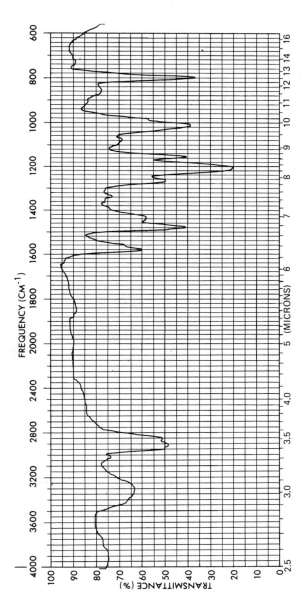

FIGURE 16.11   IR curve of DER-669 epoxy resin.

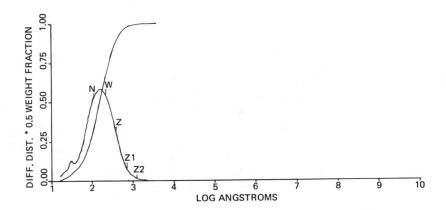

FIGURE 16.12   GPC curve of extracted epoxy resin.

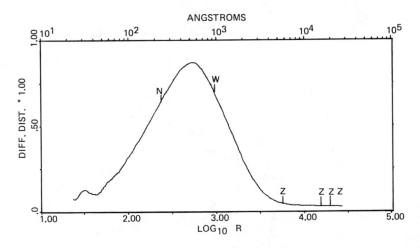

FIGURE 16.13   GPC curve of epoxy resin used before grafting.

of the reasons could be that there are more grafting sites available in a
higher molecular-weight epoxy resin. This can be demonstrated as follows:
In a high molecular weight epoxy resin (Mn ~ 10,000) there are roughly 34
repeating units of

$$-O-CH_2-\overset{\overset{\displaystyle OH}{|}}{C}H-CH_2-O-$$

in the backbone. There are, therefore, also 34 x 5 = 170 hydrogens that
can be abstracted by free radical to form a radical site where initiation of
monomers can occur.

$$-O-CH_2-\overset{\overset{\displaystyle OH}{|}}{C}H-CH_2-O- \xrightarrow{R\cdot} -O-CH_2-\overset{\overset{\displaystyle OH}{|}}{\underset{\cdot}{C}}-CH_2-O- + RH$$

$$-O-CH_2-\overset{\overset{\displaystyle OH}{|}}{\underset{\underset{\underset{\displaystyle M\cdot}{M}}{M}}{C}}-CH_2-O-$$

Now if a low molecular-weight eposy resin is used, e.g., Mn 1000,
Epon 1001 or DER 661 there are two

$$-O-CH_2-\overset{\overset{\displaystyle OH}{|}}{C}H-CH_2-O-$$

in the epoxy backbone, and only 10 abstractable hydrogens.

CHARACTERIZATION OF GRAFTING SITE

Due to the very small concentration of grafting sites, it is normally
very difficult to detect them. An attempt was made to find out what were
the most probable grafting sites on epoxy resin. [13]C nuclear magnetic
resonance (NMR) spectroscopy was used [6]. In ^{13}C NMR, experiments
were run in such a way that the carbons on the epoxy resin will be magnified
at the expense of the acrylic type carbons. Before going to the actual graft
copolymer a model compound was prepared. The model compound is

$$\text{C}_6\text{H}_5\text{-O-CH}_2\text{-}\overset{\displaystyle \text{OH}}{\overset{\displaystyle |}{\text{CH}}}\text{-CH}_2\text{-O-C}_6\text{H}_5$$

This was prepared by reacting phenol with phenyl glycidyl ether. Other model compounds such as

$$\text{C}_6\text{H}_5\text{-O-CH}_2\text{-}\overset{\displaystyle \text{O}}{\overset{\displaystyle / \backslash}{\text{CH}}}\text{-CH}_2 \qquad\qquad \text{C}_6\text{H}_5\text{-O-CH}_2\text{-}\overset{\displaystyle \text{OH}}{\overset{\displaystyle |}{\text{CH}}}\text{-CH}_2\text{-Cl},$$

and $\quad$ $\text{C}_6\text{H}_5\text{-O-CH}_2\text{-}\overset{\displaystyle \text{OH}}{\overset{\displaystyle |}{\text{CH}}}\text{-CH}_2\text{-OH}$

were also prepared, but they were considered to be of little significance due to the very small concentration of these types of structures in the epoxy resin.

The ^{13}C spectra were run in DMSO (dimethyl sulfoxide) as solvent which also acts as the internal standard. The peak positions and peak intensities of the various carbons in the model compound are listed in Table 16.11.

In order to estimate the amount of reactions at each carbon the peaks were normalized with the para-carbon of the phenyl group as unity (Table 16.12). There is a significant decrease in the peak height of the

$$-\text{C}_6\text{H}_4\text{-O-CH}_2\text{-};$$

~30% decrease. For the model compound, the grafting appears to take place at the -O-CH2- carbon.

It is surprising to see that in the model compound there is not decrease in the peak height in the carbon bearing the tertiary hydrogen (tertiary carbon). The tertiary hydrogen is probably more reactive than the secondary hydrogen, but statistically there are four times as many secondary hydrogens. Under photolytic condition, the degradation of model compounds for epoxy resin is known [7] to take place at the phenoxycarbon bond as shown in the following

$$\text{C}_6\text{H}_5\text{-O-CH}_2\text{-}\overset{\text{OH}}{\overset{|}{\text{CH}}}\text{-CH}_2\text{-O-} \xrightarrow{\text{U. V.}} \left(\text{C}_6\text{H}_5\text{-O}\cdot + \cdot\text{CH}_2\text{-}\overset{\text{OH}}{\overset{|}{\text{CH}}}\text{-CH}_2\text{-O-} \right) \longrightarrow$$

$$\text{C}_6\text{H}_5\text{-OH} + \text{CH}_2\text{=}\overset{\text{OH}}{\overset{|}{\text{C}}}\text{-CH}_2\text{-O-} \longrightarrow \text{CH}_3\text{-}\overset{\text{O}}{\overset{||}{\text{C}}}\text{-CH}_2\text{-O-}$$

TABLE 16.11
^{13}C NMR Analysis of Model Compound,
Phenyl Glycidyl Ether

Model Compound	118.247[a] 28[b]	88.734 99	80.086 51	74.211 92	29.011 55	27.371 32
Model Compound + Acrylic	118.249 44	88.786 201	80.096 98	74.229 201	29.009 93	27.327 67

[a]Peak positions downfield from DMSO, ppm
[b]Intensity of peak

TABLE 16.12
^{13}C NMR Analysis of Model Compounds

Model Compound	1	0.6	0.33
Model Compound + acrylic	1	0.46	0.33

The assignment of these peaks in the epoxy resin is shown in the following. There is a small shift from those of the model compound (Table 16.13). The peak positions reported in ppm are downfield from TMS.

Previously a series of experiments with different levels of free-radical initiator were run and it was shown that by fractional precipitation there was a straight-line relationship between the increase acrylic solubility or relative % grafting with the % free-radical initiator (Figure 16.7). These samples made with different levels of free-radical initiator were submitted for ^{13}C NMR analysis. Assuming that there is same amount of acid-epoxy reaction in these graft copolymers prepared with different levels of free-radical initiator, the peak heights at the 28.9 ppm (due to

$$\overset{\displaystyle OH}{\underset{\uparrow}{-O-CH_2-CH-CH_2)}}$$

and 27.2 ppm, due to the tertiary carbon

$$\overset{\displaystyle OH}{(-O-CH_2-\underset{\uparrow}{CH}-CH_2-O-)}$$

were plotted against the free-radical initiator concentration (Figures 16.14 and 16.15). Clearly, the peak height of both

$$\overset{\displaystyle OH}{-O-CH_2- \text{ and } -CH-}$$

decreased with increased concentration of free-radical initiator, confirming the solvent fractionation to data.

In one experiment, where grafting was carried out without solvent, there was a significant decrease in the peak height at the methylene carbon α to the epoxide group (peak at 28.2 ppm).

A plot of free-radical initiator level versus the peak height of the 28 ppm peak using one of the phenoxycarbon peak (73.6 ppm) as internal standard was made (Figure 16.16). In contrast, the peak at 28 ppm actually increased with free-radical initiator content indicating that there is probably no significant carbon-carbon bond grafting at that carbon atom. The decrease in the peak for the methylene carbon α to the epoxide group (peak at 28.2 ppm) after grafting is probably due to the following reaction.

$$\text{〈〉}-O-CH_2-\overset{O}{\overset{/\backslash}{CH-CH_2}} + HO-\overset{O}{\overset{\|}{C}}-R \longrightarrow \text{〈〉}-O-CH_2-\overset{OH}{\overset{|}{CH}}-CH_2-O-\overset{O}{\overset{\|}{C}}-R$$

TABLE 16.13
Peak Assignment in Epoxy Resin

$\overset{\displaystyle OH}{\underset{\displaystyle \uparrow}{\vert}}$ -CH-	$\overset{\displaystyle OH}{\underset{\displaystyle \uparrow}{\vert}}$ -CH₂-CH-	-CH₂-CH-CH₂ $\longrightarrow$ ⬡-	
PPM	27.180	28.894	28.228	73.501

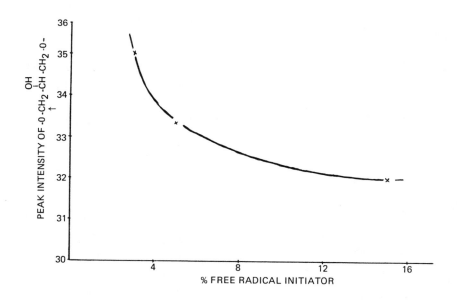

FIGURE 16.14 Peak intensity of the $-O-CH_2-\overset{OH}{\underset{}{\vert}}{CH}-CH_2-O-$ peak versus free radical initiator concentration (from ^{13}C NMR data).

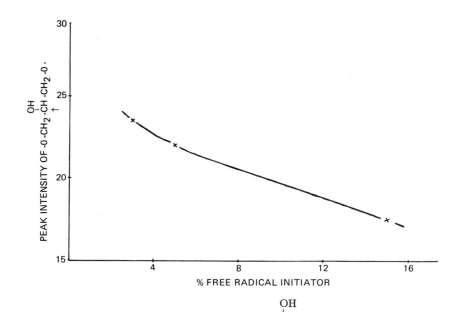

FIGURE 16.15 Peak intensity of the $-O-CH_2-\overset{\underset{\displaystyle OH}{|}}{C}H-CH_2-O-$ peak versus free radical initiator concentration (from ^{13}C NMR data).

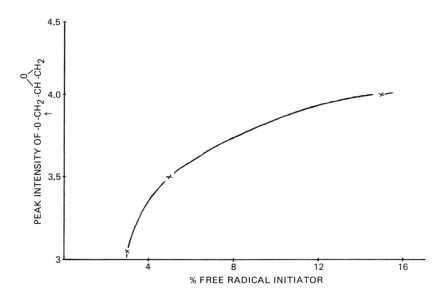

FIGURE 16.16 Peak intensity of $-O-CH_2-CH-CH_2$ peak versus free radical initiator concentration (from ^{13}C NMR data).

The ^{13}C NMR spectra of the epoxy resin and the graft copolymer are shown in Figures 16.17 and 16.18.

MECHANISM OF GRAFTING

From a classical polymerization scheme, there is a) initiation, b) propagation, c) chain transfer, and d) termination.

In an idealized case, where the transfer mechanism only occurs to a "Foreign Polymer," i.e., to say transfer to monomer is negligible, an equation can be derived to show that the rate of grafting is:

$$V_r = k_r \, [Pr] \, [R] \, [M] \tag{1}$$

where [Pr] is the concentration of growing polymer chains formed from the initiator; [R] is the transfer polymer added to the system; and [M] is the concentration of monomer.

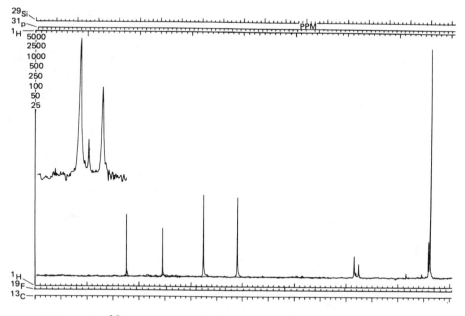

FIGURE 16.17   ^{13}C NMR spectrum of epoxy resin.

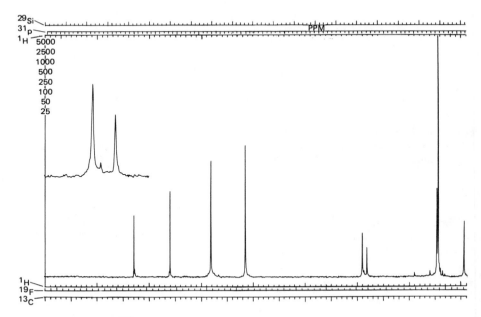

FIGURE 16.18 ^{13}C NMR spectrum of graft copolymer.

Therefore, the highest yields of grafted copolymer should be obtained under the following conditions: 1) increasing concentration of transfer polymer to a limiting value; and 2) high rates of initiation obtained by a) increasing initiator concentration, and b) increasing the polymerization temperature.

In practice, it is not usually possible to take advantage of these conditions since the majority of polymers have limited solubility in "foreign" monomer solutions, particularly at elevated temperatures where the high initiation rates lead to gelation and phase separation. However, substitution of a resin of low molecular-weight for the conventional backbone polymer of high molecular-weight enables these difficulties to be overcome to a large extent. "Resins" of low molecular weight such as conventional epoxy resins, polyethylene glycols, poly(methoxy acetals), etc., are readily soluble in such monomers as methyl methacrylate, styrene, vinyl acetate, etc., to give high concentrations of "grafting polymer." Even at comparatively high rates of initiation these systems in many cases remain completely compatible and polymerization can be taken to complete conversion without phase separation on a macroscale.

There are numerous examples in the literature [1, 8, 9, 10] where free radicals are generated on the polymer backbone followed by grafting of monomer onto the polymer, i.e., initiation of monomer occurring after chain transfer reaction. For example [8], in grafting monomer onto polyester fiber, active centers seem to be created by direct hydrogen abstraction from the polyester molecules by the primary free-radical species benzoyloxy radical or by the secondary free-radical species phenyl radical. Benzol peroxide (BPO) was used as initiator. There was no mention of benzoic acid or benzene formation. The other mechanism mentioned was by oxidizing the polyester to hydroperoxide at several points along the chain in a random manner. The hydroperoxide decomposes into the active form at high temperature to produce ultimately macroradicals one of which may be represented as

$$-C \overset{\overset{\text{O}}{\|}}{} - \langle \bigcirc \rangle - \overset{\overset{\text{O}}{\|}}{C} - O - CH - CH_2 -$$

These readical sites permit attachment of monomer molecules which may grow into short chains. This is an example of "grafting from" [2] process where the free radical (or other active site) is generated on the backbone and subsequently it initiates the polymerization of monomers to produce branches. The other grafting process is the "grafting onto" where a growing free radical (or other active species) attacks another preformed polymer preferentially carrying suitable substituents and thereby produce a branch on the preformed backbone.

Graft epoxy-acrylic copolymer prepared with free-radical initiator is an example of the "grafting from" process. In the case where benzoyl peroxide was used as the free-radical initiator, it is determined that about 77% of the free-radical initiator instead of causing initiation of monomers, chain transfer with the epoxy resin backbone, followed by the "grafting from" of monomers onto the epoxy resin. Benzoyl peroxide is known [11] to decompose mostly (90%) to the benzoyloxy radical and (10%) phenyl radical. Mechanisms of grafting can be demonstrated in the two schemes (A and B).

There is probably a very small amount of Scheme B present as the aliphatic protons of the epoxy resin are much more activated towards free radical abstraction than the hydrogen of the 2-butoxy-ethanol-1. The 23% of the benzoyl peroxide that did not chain transfer would initiate polymerization of monomers to form ungrafted styrene-methacrylic acid copolymer. This could be part of the reason that the ungrafted or free styrene-methacrylic acid copolymer is of higher molecular weight than that of the styrene-methacrylic acid copolymer made under the same conditions in the absence of the epoxy resin. This lower amount of free-radical initiator would result in a higher molecular-weight copolymer. The other reason is possibly due to a viscosity effect, such as the Trommsdorf effect, where higher molecular polymer is obtained due to lesser chance of termination

SCHEME A

Transfer with Epoxy resin backbone initially

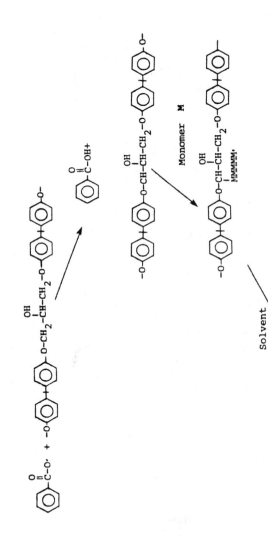

Monomer M

Solvent

Termination by chain transfer
with solvent or epoxy resin

The latter is probably more likely due to proximity and concentration of the aliphatic
hydrogens of the epoxy backbone.

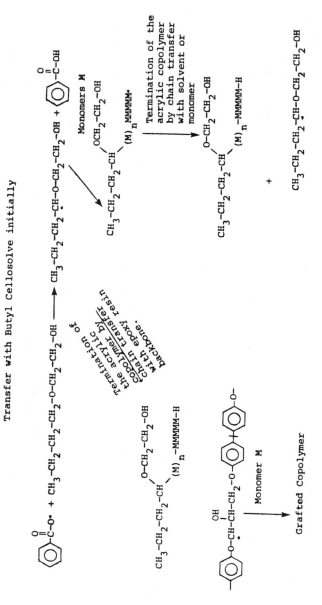

SCHEME B

Transfer with Butyl Cellosolve initially

in a more viscous medium (i.e., in the presence of high molecular-weight epoxy resin).

From the data generated so far, an attempt was made to determine the epoxy-acrylic graft copolymer composition.

Seventy-seven percent of the benzoyl peroxide formed benzoic acid by hydrogen abstraction (from Example III). $11.6 \times 0.77 = 9$ mmol of BPO is involved in grafting.

Since about half of the epoxy resin and about two-thirds of the acrylic are free, the grafted composition is as listed in Table 16.14. Therefore, per epoxy molecules, there are about two grafting sites, and each grafted chain consists of ~9 acrylic units. The composition of the acrylic units are roughly 2:1 methacrylic acid; styrene or

x here is roughly equal to 3.

The composition of the grafted copolymer is thus shown in the following:

TABLE 16.14
Epoxy-Acrylic Graft Copolymer Composition

	Starting composition mmol	Graft composition mmol	Ratio
Epoxy	10	5	1
Acrylic	218	87	17
Benzoyl peroxide	11.6	9	2

Percent grafting efficiency can be calculated from

$$\% \text{ grafting efficiency} = \frac{\text{monomer grafted}}{\text{monomer grafted} + \text{free copolymer}}$$

$$= \frac{8}{20} = 40\% \tag{2}$$

The apparent degree of grafting [12] is defined as

$$G = \frac{W-W_o}{W_o} \times 100 \tag{3}$$

where G is the apparent degree of grafting; W_o the weight of sample before grafting; and W the weight of sample after grafting.

For the epoxy-acrylic graft copolymer, the apparent degree of grafting is

$$G = \frac{(\text{Free epoxy} + \text{graft epoxy-acrylic copolymer}) - \text{epoxy before grafting}}{\text{Epoxy before grafting}} \times 100 \tag{4}$$

$$= \frac{88-80}{80} \times 100 = 10\%$$

Anchor/graft ratio based on molecular weight is A/G = 8858/828 = 10.7.

SUMMARY

The synthesis of an epoxy-acrylic graft copolymer using free-radical means was described. Characterization of the graft copolymer by solvent extraction indicated that the graft copolymer consists of the following: 1) 47% of the epoxy resin is ungrafted; 2) 61% of the acrylic monomer polymerizes to form acrylic copolymer; 3) 39% of the acrylic monomer is grafted onto 53% of the epoxy resin.

Characterization of grafting sites on the epoxy resin by NMR spectroscopy indicated that grafting takes place at the aliphatic carbon atoms of the epoxy resin.

The mechanism of grafting was discussed. The mechanism follows the "grafting from" process in which a free radical is generated on the epoxy backbone and subsequently it initiates the polymerization of monomers to produce branches. The composition of the epoxy-acrylic graft polymer is also discussed. For each epoxy resin molecule, ~ 8000 molecular weight, there are about two grafting sites. Each grafted acrylic chain consists of ~ 828 molecular weight.

ACKNOWLEDGMENT

The author wishes to express his sincere thanks to Dr. V. Ting and Dr. J. Evans for their contribution to this work. The authors also wish to thank Glidden Co., Division of SCM Corporation for permission to publish this work.

REFERENCES

1.   R. J. Ceresa, The Chemistry of Polymerization Processes, A.C.S. Monograph #20, p. 249, 1966.
2.   J. P. Kennedy, Polym. Symp. 64:117, 1978.
3.   G. Mino and S. Kaizerman, J. Polym. Sci., 31:242, 1958.
4.   R. J. Ceresa, J. Polym. Sci. 53:9, 1961.
5.   N. Ohto, E. Niki and Y. Kamiya, J. Chem. Soc., 1416, 1976.
6.   S. A. Sojka and W. B. Monita, J. Appl. Polym. Sci., 20:1977, 1976.
7.   P. Pappas. Private communication.
8.   S. H. Abdel-Fattah, S. E. Shalaby, E. A. Allam and A. Habeish, J. Appl. Polym. Sci., 21:3355, 1977.
9.   B. N. Misra and R. Dogra, J. Macromol. Sci. Chem., A14(5):763, 1980.
10. C. G. Beddows, H. Gil and J. T. Guthrie, Polym. Bull. 3:645, 1980.
11. P. W. Allen, G. Ayrey and C. G. Moore, J. Polym. Sci., 36:55, 1959.
12. F. Sundardi, J. Appl. Polym. Sci. 22:3163, 1978.

NEW STUDIES OF THE DE-ADHESION OF COATINGS FROM METAL SUBSTRATES IN AQUEOUS MEDIA: INTERPRETATION OF CATHODIC DELAMINATION IN TERMS OF CATHODIC POLARIZATION CURVES

Henry Leidheiser, Jr., Lars Igetoft,
Wendy Wang, and Keith Weber

Center for Surface and Coatings Research
Lehigh University
Bethlehem, Pennsylvania

ABSTRACT

Cathodic delamination studies in aerated 0.5 M NaCl were performed on 15 μm thick polybutadiene coatings on copper and tin substrates and 110 μm thick alkyd coatings on a steel substrate. Cathodic polarization curves were determined for copper, tin and steel in 0.5 M NaCl and on steel in 0.5 M NaCl adjusted to pH 10 and 12.5. Comparison of these two types of data allowed the following two conclusions to be drawn: (1) The rate of delamination of polybutadiene coatings from copper and tin at moderate cathodic overvoltages is not controlled by lateral diffusion of OH$^-$ ions generated at the defect by the cathodic reaction. (2) The rate of delamination of alkyd coatings on steel increases greatly when the cathodic potential is sufficient for the generation of hydrogen by the reaction, $2H^+ + 2e^- = H_2$.

INTRODUCTION

An understanding of the delamination of organic coatings from metal substrates upon making the system the cathode while immersed in an electrolyte is a continuing interest in our laboratory. Previous work has been concerned with outlining some of the important variables in the delamination process [1,2], a summary of the various mechanisms by which

307

delamination may occur [3], and a summary of the methods by which delamination may be controlled [4]. The prior work has emphasized the
importance of the electrochemical nature of the delamination process. The
work described herein gives several examples of the application of cathodic
polarization curves to the interpretation of the delamination phenomenon.

Three systems will be discussed: polybutadiene coatings on copper
substrates; polybutadiene coatings on tin substrates; and a commercial
alkyd coating on a steel substrate. Each of these systems has been selected
to illustrate a specific point.

EXPERIMENTAL

Three substrates were used: copper sheet, commercial cold rolled
steel and commercial tinplate. The copper and the steel panels were
abraded with 240 grit abrasive paper while wetted with water, washed in
tap water and finally rinsed with acetone. The tin panels were polished
with aluminum oxide powder and a piece of cloth wetted in water and, there-
after, washed and rinsed as the other panels.

Prepared panels were kept in a desiccator until application of the
coating. All paint layers were applied by spinning in a centrifugal applicator. Two different coating systems were used: unpigmented polybutadiene
and a commercial pigmented alkyd. The polybutadiene was Budium RK-662
from DuPont applied in one layer. It was cured at 200°C in an air oven for
20 min. The coating thickness, determined by means of a micrometer, was
15 μm in all cases.

The alkyd system was applied in the form of two layers of primer and
two layers of top coat. Each layer was 25-30 μm in thickness and the total
thickness of the dry coating was 105-110 μm. A drying time of 24 h was
used between each coat and the 4-coat system was dried at room temperature for not less than 2 weeks before immersion in electrolyte. The composition of the primer (Rust-Oleum 7773) and the topcoat (Rust-Oleum satin
white 7791) were as follows:

Primer
   Pigment   44.6%
      Silicates   15.6%
      Yellow iron oxide   7.6%
      TiO_2 2.7%
      Calcium borosilicate 18.2%
      Bentonite clay   0.5%
   Vehicle   55.4%
      Linseed phenolic alkyd resin 20.1%
      Aliphatic hydrocarbons   32.6%
      Driers and additives   2.7%

Topcoat
    Pigment   53.5%
        Silicates   34.7%
        TiO_2 16.7%
        Bentonite clay   0.1%
        Tinting colors   2.0%
    Vehicle   46.5%
        Soya/menhaden alkyd resin   17.2%
        Aliphatic hydrocarbons   28.8%
        Driers and additives   0.5%

The edges and the back sides of the samples, 2.3 X 5 cm, were protected with a thick commercial epoxy coating so as to limit exposure of the area under study to 2-4 cm^2. Before immersion in the electrolyte, an indentation was made with a sharp steel needle in the center of the sample area in order to form a circular defect of approximately 0.01 cm^2.

The potential of the sample during the experiment was maintained constant at a predetermined value by means of a potentiostat. The potential of the cathode was sensed with a Luggin capillary whose tip was maintained near the defect. The counter electrode was graphite. All potentials were determined with reference to a saturated calomel electrode (SCE) and are reported in terms of this reference.

The delaminated area of the coatings at the end of the experiment was determined by pressing adhesive tape against the coating and measuring the area that separated from the metal. The area adjoining the delaminated region remained tightly adherent as judged by probing with a needle.

Cathodic polarization curves were determined by the procedure outlined in ASTM standard G5-72. The data were obtained in the potentiodynamic mode at a scan rate of 0.33 mv/sec. Measurements were made using a Princeton Applied Research (PAR) Model 350 Corrosion Measurement Console and a Model K47 Corrosion Cell system. The counter electrode was graphite.

All measurements were made in 0.5 M NaCl solution open to the air.

RESULTS

Polybutadiene Coatings on Copper

The concept for the experiment which is described below originated as a result of the generation of cathodic polarization curves for about 10 metals in aerated 0.5 M NaCl solution. The cathodic polarization curve for copper was of special interest because the cathodic current was approximately constant over a range of -0.4 to -1.1 V as shown in Figure 17.1. The constant current region at approximately 3 X 10^4 namp/cm^2 over this

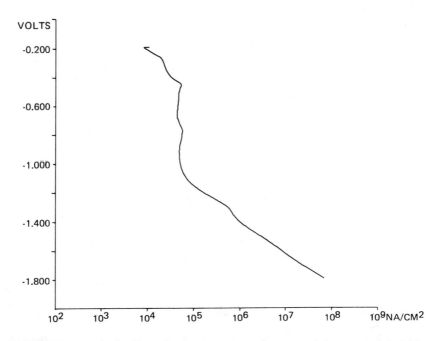

FIGURE 17.1   Cathodic polarization curve for copper in aerated 0.5 M
                NaCl.

potential range is a consequence that the rate limiting step is the rate of
diffusion of oxygen to the surface. The curve is a direct reproduction of
the printed curve from the PAR 350 Corrosion Console Measurement system and contains a minor anomaly caused by a scale change within the data
accumulation electronics. A similar polarization curve is obtained for
copper coated with 15 μm of polybutadiene and then damaged with a pointed
instrument to expose a copper area of approximately 0.01 cm^2 as shown in
Figure 17.2. No significant delamination occurred during the time required
to generate the polarization curve. The similarity of these curves indicates that the cathodic behavior of a defect in the coated metal is the same
as that of uncoated copper.

     Delamination studies were then carried out at a cathode potential of
-0.6, -0.8 and -1.0 V on polybutadiene-coated copper with a coating thickness of 15 μm. Each of the datum points in Figures 17.3 and 17.4 represents a different experiment, in all of which the current passing through the
system during the entire experiment ranged from 0.2 to 0.4 μamp after
steady-state conditions were obtained. Figure 17.3 gives the delamination
as a function of time and Figure 17.4 gives the delamination as a function
of the number of coulombs passed through the system.

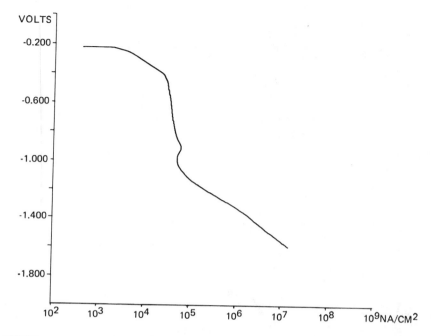

FIGURE 17.2   Cathodic polarization curve obtained for a defect, 0.01 cm^2, in copper coated with polybutadiene, 15 μm in thickness.

Polybutadiene Coatings on Tin

Similar experiments were carried out also with polybutadiene-coated tinplate. Figure 17.5 shows the polarization curve for tin in aerated 0.5 M NaCl. The peak at approximately -1.0 V presumably represents additional consumption of electrons by reduction of tin oxide. Cathodic delamination measurements were performed at -0.8, -1.0, and -1.2 V and the results are shown in Figures 17.6 and 17.7. The current passing through the defect at -1.2 and -1.0 V was approximately 0.2 μamp and at -0.8 V it was less than 0.1 μamp, values which are in reasonable accord with the polarization curve in Figure 17.5 when account is taken that the surface area of the defect is approximately 0.01 cm^2.

Alkyd Coatings on Steel

The delamination as a function of time is given in Figure 17.8 for cathodic potentials of -0.8, -1.0, -1.2, -1.4, -1.6, and -1.8 V. It will be noted that the delamination achieved a linear value with time after a

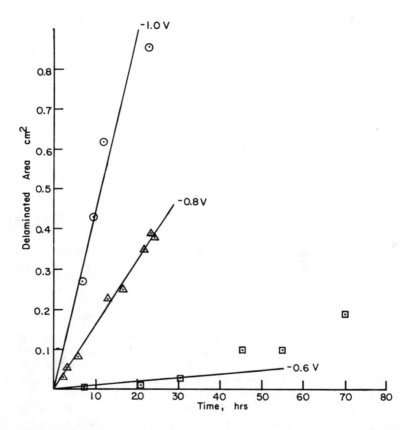

FIGURE 17.3   Delamination as a function of time for 15 μm thick polybutadiene coatings on copper immersed in 0.5 M NaCl when
cathodically polarized to -0.6, -0.8, and -1.0 V.

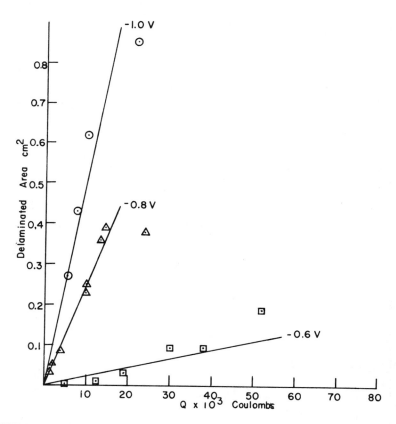

FIGURE 17.4   Delamination as a function of coulombs for 15 μm thick
polybutadiene coatings on copper immersed in 0.5 M NaCl
when cathodically polarized to -0.6, -0.8, and -1.0 V.

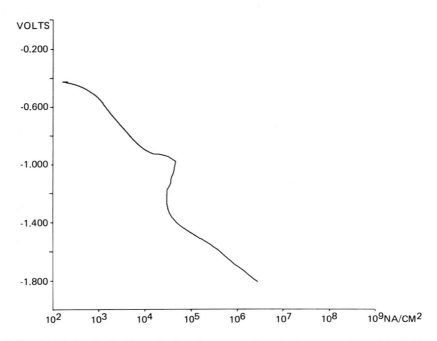

FIGURE 17.5   Cathodic polarization curve for tin in aerated 0.5 M NaCl.

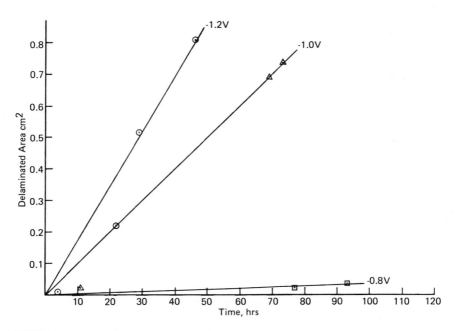

FIGURE 17.6   Delamination as a function of time for 15 μm thick polybutadiene coatings on tin immersed in 0.5 M NaCl when
cathodically polarized to -0.8, -1.0, and -1.2 V.

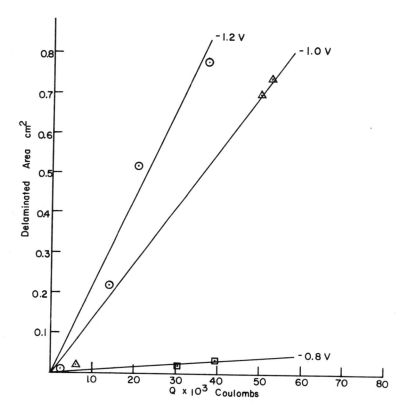

FIGURE 17.7   Delamination as a function of coulombs for 15 μm thick
polybutadiene coatings on tin immersed in 0.5 M NaCl
when cathodically polarized to -0.8, -1.0, and -1.2 V.

period of time. This behavior is in sharp contrast to the delamination
behavior of polybutadiene coatings, 15 μm in thickness, on copper (see
Figure 17.3) where the rate was linear with time from the very beginning of
the experiment. This time to achieve a linear rate is termed "delay time"
and is obtained by extrapolation of the linear portion of the curve to zero
delamination. Values obtained by such extrapolation are given in Table
17.1 where is is apparent, with one exception, that the delay time decreased
with increase in the cathode potential. The delamination rate in the linear
portion of the delamination curve is plotted in Figure 17.9 as a function of
the cathode potential.

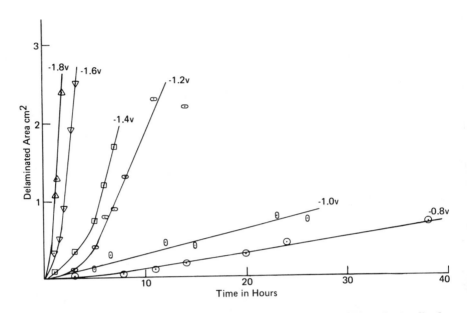

FIGURE 17.8   Delamination as a function of time for 110 μm thick alkyd
coatings on steel immersed in 0.5 M NaCl when cathodically
polarized to -0.8, -1.0, -1.2, -1.4, -1.6, and -1.8 V.

TABLE 17.1
A Summary of the Delay Times in Achieving a Linear
Delamination Rate in the Case of 110 μm Thick
Alkyd Coatings on Steel Cathodically Treated
in 0.5 M NaCl

Potential	Delay Time Obtained by Extrapolation of Linear Rate to Zero Delamination (h)
-0.8 V	6 h
-1.0	2 [a]
-1.2	4
-1.4	3
-1.6	1.5
-1.8	ca 0.7

[a]Time difficult to determine because of scatter in data.

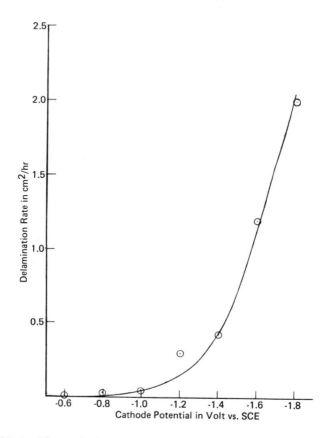

FIGURE 17.9   The cathodic delamination rate of 110 μm thick alkyd
coatings on steel immersed in 0.5 M NaCl as a function
of potential.

It is well known that the liquid beneath the coating in the delaminated
area has a high pH [6]. It thus was appropriate to determine the cathodic
polarization curve as a function of pH in order to have information about
how the cathodic polarization curve might change within the liquid beneath
the coating. Such cathodic polarization curves are given in Figures 17.10,
17.11, and 17.12 for pH values of 6.5, 10, and 12.5 at the start of the experiment. The pH at the completion of the experiment was within 0.3 pH
units of that at the start in the case of the alkaline solutions.

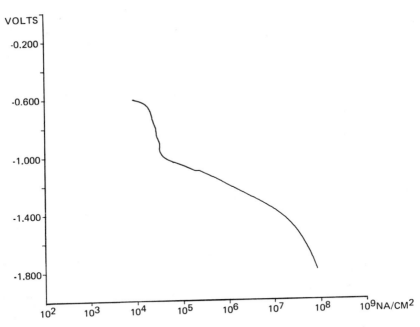

FIGURE 17.10   The cathodic polarization curve for steel in aerated 0.5 M
                NaCl at pH 6.5.

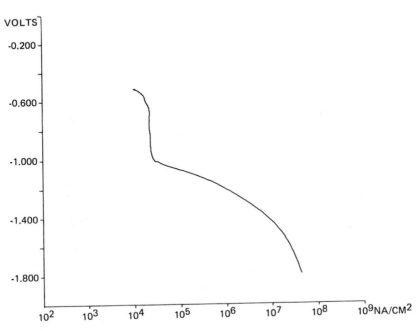

FIGURE 17.11   The cathodic polarization curve for steel in aerated 0.5 M
                NaCl at pH 10.

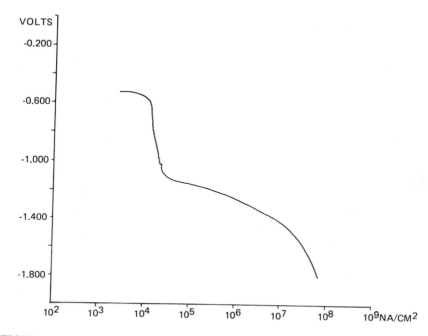

FIGURE 17.12 The cathodic polarization curve for steel in aerated 0.5 M
NaCl at pH 12.5.

DISCUSSION

The cathodic polarization curves for copper and steel in aerated 0.5
M NaCl have the same general shape, a region where the current density
is roughly constant with increase in driving force for the cathodic reaction,
and a region where the logarithm of the current increases linearly with
potential. The first region extends from approximately -0.4 V to approximately -1.1 V in the case of copper and from approximately -0.6 V to
approximately -1.1 V in the case of steel. In the case of tin this insensitivity to applied potential extends from approximately -1.0 V to approximately -1.4 V but is complicated by an oxide reduction peak at approximately -1.0 V. This potential-independent region is associated with the
cathodic reaction, $H_2O + 1/2 O_2 + 2e^- = 2 OH^-$, and the insensitivity to
potential is a consequence of the fact that the rate is limited by the rate at
which oxygen can reach the interface and participate in the cathodic reaction. At more cathodic potentials, in the region where the logarithm of
the current is linearly related to potential, the competing reaction, $2H^+ +
2e^- = H_2$, begins to become dominant and the majority of the current is
consumed in supporting this latter reaction. Gas bubbles begin to become

visible at the defect when the current exceeds 10^5 namp/cm^2 as is expected
on the basis of the polarization curves. Thus, the polarization curve provides information that aids in determining the reaction which takes place at
the defect when the cathode potential is known.

The experiments with polybutadiene-coated copper and tin were all
carried out at potentials where the major electron-consuming reaction at
the defect is $H_2O + 1/2\ O_2 + 2e^- = 2\ OH^-$. The potential of the metal
radially from the defect is less negative because of resistive components
and thus the same cathodic reaction likely occurs under the coating as at
the defect. The absence of any significant hydrogen evolution reaction
under the coating is supported by the fact that no gas bubbles were observed
under the coating or emerging from the defect so long as the potential was
maintained at a less negative value than the point at which the logarithm of
the current becomes a linear function of the cathode potential.

It has been noted previously that there is a linear relationship between
the delamination as a function of time (delamination rate) and the delamination as a function of the number of coulombs passed (delamination parameter) in the case of thin polybutadiene coatings on steel. This same relationship is obeyed for thin polybutadiene coatings on copper and tin with the
points for all metals falling on the same line. In these three cases, such a
relationship exists because the current is approximately the same and there
is little change in the current during the term of the experiment. The
cathodic polarization curves in aerated 0.5 M NaCl for copper (Figure 17.1),
for tin (Figure 17.5), and for iron (Figure 17.10) are approximately the same
over the potential range studied and thus the current flowing through the
defect is the same, namely, 3×10^4 namp/cm^2. The similarity of the
current is a result of the fact that the rate limiting step for the current
density at the defect is the rate of diffusion of oxygen to the surface.

The experiments carried out with copper at potentials of -0.6, -0.8,
and -1.0 V and those with tin at -1.0 and -1.2 V all involved cathodic cur-
rents of the order of 0.3 μ amp during the entire experiment, yet the delamination parameter and delamination rate were very different at the different potentials as shown by the data summarized in Table 17.2. The similarity of the current and the fact that the experiments were carried out at
potentials where oxygen access to the electrode surface is rate controlling,
as shown by the polarization curves, indicate that the number of hydroxyl
ions generated per unit time at the defect was similar in all the experiments summarized in Table 17.2. However, the delamination rate and the
delamination efficiency, as measured by the delamination parameter, were
widely different. These data thus provide convincing proof that the major
delaminating mechanism cannot be lateral diffusion of hydroxyl ions away
from the defect and under the coating in the case of thin polybutadiene
coatings on copper, tine, and steel.

The cathodic polarization curves for steel in 0.5 M NaCl at pH's of
6.5, 10, and 12.5 are shown in Figures 17.10-17.12. The shapes of the curves
are similar although they are shifted slightly to the left with increasing pH.

TABLE 2
Delamination Rates and Delamination Parameters for Copper
and Tin in Those Experiments in Which the Current Passing
through the Interface Was the Same

	Copper		Tin	
Potential	Delamination Rate in cm^2/h	Delamination parameter in $cm^2/coulomb$	Delamination Rate in cm^2/h	Delamination parameter in $cm^2/coulomb$
-0.6	0.001	2.3		
-0.8	0.016	25		
-1.0	0.045	52	0.0095	13.5
-1.2			0.017	22.3

The significance of these curves is that the cathodic behavior of a steel surface over a pH range of 6.5 to 12.5 is approximately the same. Rationalization of effects on the basis of the cathodic polarization curves can thus be applied on the same basis for the metal surface at the delaminating edge and at the defect.

The delamination data obtained with the 110 μm thick alkyd coatings on steel shown in Figure 17.9 indicate a large increase in delamination rate when the cathode potential is more negative than -1.1 V. Comparison between Figures 17.9 and 17.10 shows that this increase in rate coincides with the onset of the hydrogen evolution reaction at the defect. We interpret these results on the basis that the hydrogen evolution reaction at the defect creates a stress at the coating/substrate interface and a mechanical component is thus added to the chemical component of the delaminating process. Some support for this concept is found in the plot given in Figure 17.13 in which the delamination rate for the alkyd coating in the potential region where hydrogen evolution occurs is a function of the cathodic current density at the defect as calculated from the polarization curve.

The delamination data obtained with the relatively thick alkyd coatings differ from the delamination data obtained with the relatively thin polybutadiene coatings on various substrates in that a delay time is required before the delamination increases linearly with time. These delay times decrease as the cathode potential becomes more negative (Table 17.1). The delay time is explained in terms of the time required to establish a steady state condition for a diffusion gradient in the coating. It is not known if the rate controlling factor in establishing the gradient is oxygen, water, or

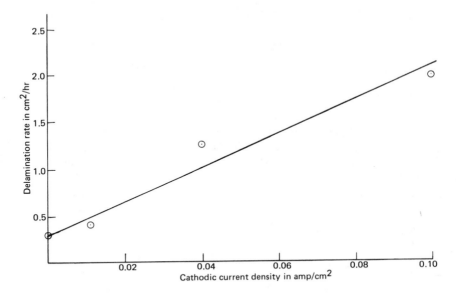

FIGURE 17.13   The delamination rate of 110 μm thick alkyd coatings on
                steel as a function of the current density at the defect as
                estimated from the polarization curve.

cations. The steady state condition takes longer to establish when the potential gradient across the coating is low (less negative cathode potential)
than when the potential gradient is high (more negative cathode potential).
        The view has previously been expressed [1] that the rate controlling
step in the delamination process is the rate at which cations move through
the coating to provide counterions for the hydroxyl ions generated by the
cathodic reaction. The high pH of the liquid under the coating is proof that
hydrogen ions which may diffuse through the coating are not serving as
counterions. Dickie [5] has recently revealed that sodium ions do concentrate at the metal/organic coating interface in salt spray tests of scribed
panels where the delaminated region is cathodic to the anodic sites at the
scribe marks. None of the experiments done to date provides evidence
that allows a distinction to be made between sodium cations that diffuse
through the coating and sodium cations that reach the delaminating edge
from the liquid in the delaminated region. The decrease in delaminating
rate with increase in coating thickness [1] and the effect of potential on the
delay time for the alkyd coatings are suggestive that cation diffusion through
the coating is a critical factor, but they are not proof of the fact. Studies
are now underway to determine the rate of cation transport through the
coating and the role of the cation in charge transport through the coating,
but such data are not yet available.

An estimate of the efficiency of the delaminating process in the case of copper can be made if it is assumed that every electron that is involved in the cathodic reaction under the coating leads to the breaking of a single bond and that there are 10^{15} bonds/cm^2 of interfacial area, i.e., the number of atoms/cm^2 of substrate assuming a roughness factor of 1. The efficiency is then calculated from the relationship:

$$\frac{(\text{area delaminated in cm}^2) \times (1.6 \times 10^{-4} \text{ coulombs/cm}^2)}{(\text{total no. of coulombs passed through interface})}$$

On this basis the delamination efficiency is 0.018% at -0.6 V, 0.3% at -0.8 V, and 0.8% at -1.0 V.

These data allow a rough estimate of the resistance of the coating. If it is assumed that the delaminating efficiency of 0.8% at -1.0 V represents the fraction of the current that is responsible for the cathodic reaction under the coating, the current density at the delaminating edge may be calculated from the relationship

$$\begin{pmatrix} \text{current density at delaminating} \\ \text{edge in amp/cm}^2 \end{pmatrix} = \begin{pmatrix} \text{current density at defect in} \\ \text{amp/cm}^2 \end{pmatrix}$$

$$\times \text{ (delaminating efficiency)}$$

For the measurements at -1.0 V, this current density calculates to be 1.6×10^{-7} amp/cm^2 (1.6×10^2 namp/cm^2). If all the potential drop is assumed to be across the coating, i.e., the difference between the corrosion potential and the cathode potential, a value of 0.8 V may be estimated. Applying Ohm's law, the resistance of the coating is estimated to be equivalent to 5×10^6 ohms per cm^2 for the 15 μm thick coating of polybutadiene, a value approximately the same as that experimentally determined [6].

ACKNOWLEDGMENT

This research was supported by the Office of Naval Research as part of a program concerned with corrosion control by organic coatings.

REFERENCES

1. H. Leidheiser, Jr., and W. Wang, J. Coatings Technol. 53, 77 (1981).
2. H. Leidheiser, Jr., and W. Wang, "Corrosion Control by Organic Coatings," H. Leidheiser, Jr., ed., Natl. Assoc. Corrosion Engrs., Houston, Texas, 1981, pp. 70-77.
3. H. Leidheiser, Jr., Croat. Chem. Acta 53, 197 (1980).

4. H. Leidheiser, Jr., "Solid State Chemistry as Applied to Cathodically-Driven or Corrosion-Induced Delamination of Organic Coatings," Paper No. 81, Corrosion 81, Natl. Assoc. Corrosion Engrs., Toronto, Canada, April 1981.
5. R. Dickie, paper presented at Corrosion Research Conference, Natl. Assoc. Corrosion Engrs., Toronto, Canada, April 1981.
6. H. Leidheiser, Jr. and M. W. Kendig, Corrosion 32, 69 (1976).

MECHANICS OF ELECTROSTATIC ATOMIZATION, TRANSPORT, AND DEPOSITION OF COATINGS

Guy C. Bell, Jr. and Jerome Hochberg*

E. I. du Pont de Nemours & Co., Inc.
Fabrics and Finishes Department
Marshall Laboratory
Philadelphia, Pennsylvania

ABSTRACT

The mechanics of electrostatic atomization, transport, and deposition of coatings were studied. Emphasis was given to electrostatic application of automotive finishes by a high-speed rotary atomizer. The dynamics of atomization and particle deposition were examined using high-speed video recording techniques. Spray droplet sizes and speeds were measured with laser light-scattering and Doppler methods. Charge/mass ratios were determined. The droplet parameters were related to application conditions (electrostatic voltage, fluid flow rate, mechanical forces) and to paint properties (electrical resistivity, surface tension, viscosity).

INTRODUCTION

The electrostatic spray application of paints has been a widely used industrial process for many years. Recently it has received increasing attention because of its spray efficiency benefits, and the attendant help it may provide in meeting air quality standards. The spray efficiency results from the electrostatic charge on the spray droplets, which, along with reduced air flows, causes most of the paint to deposit on the workpiece that is being painted.

---

*Experimental Station, Wilmington, Deleware.

Commercial electrostatic painting equipment is available in many designs, including hydraulic ("airless") and pneumatic (compressed air) spray guns, as well as centrifugal atomizers (rotating disks and bells). The recent evolution of high-speed rotary atomizers has received increasing industrial attention [1]. While this report applies especially to the high-speed turbine bell, the same principles also apply to other electrostatic spray methods.

The literature relating to electrostatic atomization, particle charging and deposition is extensive. The fields of application include such diverse technologies as electrostatic precipitation, high-speed ink-jet printing, ionic rocket propulsion, electrostatic imaging (xerography), and atmospheric electrostatics, as well as electrostatic spray painting. "Modern" scientific studies have existed before the well-known interests of Benjamin Franklin more than two hundred years ago.

The principles and techniques of electrostatic spray painting have been reviewed by Miller [2]. Hines [3] analyzed electrostatic painting in studies involving single jets and a slow-speed rotating-cone atomizer. The fundamentals of electrostatic powder coating were studied by Wu [4] and Golovoy [5]. Anestos, Sickles and Tepper [6] investigated the charge to mass distributions in electrostatic sprays from a compressed-air electrostatic gun. The same authors also studied the surface charge buildup during electrostatic transport to painted substrates [7]. A model for a general electrostatic spray theory has been presented by Kelly [8,9]. Extensive work on the breakup of charged jets and particle charging has been carried out by Hendricks and co-workers [10,11].

EQUIPMENT AND TECHNIQUES

The present study has encompassed the mechanics of electrostatic atomization, the characteristics of the spray cloud and the deposition of the coating on the substrate. The spray application of only liquid coating systems has been examined, with emphasis on high-speed rotary electrostatic atomizers. The equipment parameters, paint properties and performance factors included in the present study are listed in Table 18.1.

A generalized diagram of the application equipment is shown in Figure 18.1. In Zone 1 of the figure, paint in a grounded metal container is pumped at a controlled flow rate to the bell-shaped atomizer. The bell is rotated at high speed by an air turbine. The turbobell assembly, insulated from ground, is charged from a high voltage power supply creating an electrostatic field between the atomizer and the ground plane or articles to be painted. In Zone 2 the paint flows over the interior surface of the spinning bell to the edge, where atomization occurs (Zone 3). The paint droplets then travel through the air (Zone 4) to the grounded substrate on which the coating is formed (Zone 5). The dynamics of the processes which occur in this sequence of application events will be discussed in more detail.

TABLE 18.1
Electrostatic Spraying with High-Speed Rotary Atomizers

Equipment Parameters
    Turbine speed
    Shaping air pressure
    Paint flow rate
    Voltage
    Distance from substrate
        Substrate size and shape
    Atomizer design

Paint Properties
    Viscosity
    Surface tension
    Electrical resistivity
    Dielectric constant
    Evaporation rate

Performance Factors
    Atomization
    Spray pattern
        Particle size distribution
    Charge
        Current to substrate
        Particle charge per mass

    The rotary atomizer used in our studies was the Ransburg[*] turbobell. The assembly diagram is shown in Figure 18.2. High speed photographic and video techniques were used to examine the atomization and deposition processes during electrostatic spraying. Figure 18.3 shows the arrangement for observing the edge of the bell and adjacent spray space (atomization zone). Particle deposition and film formation on a substrate were studied with the viewing arrangement shown in Figure 18.4.

    Several techniques were used to characterize the spray. Spray particle size was measured within the spray cloud during spraying by means of a laser light-scattering method. The instrumentation provided the mass mean diameter and the size distribution. Spray particles in

---

[*] Ransburg Corporation, Indianapolis, Indiana 46208, U.S.A.

GENERALIZED EQUIPMENT CONFIGURATION

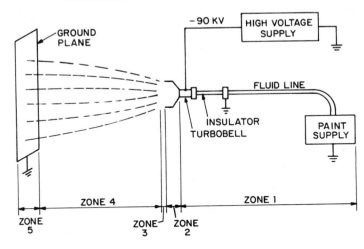

FIGURE 18.1   Generalized diagram of the application equipment.

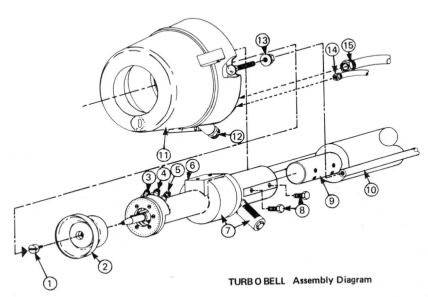

TURBO BELL Assembly Diagram

FIGURE 18.2   Assembly diagram for the turbobell (Ransburg Corporation):
(1) hub retainer screw, (2) bell assembly, (3) paint feed
tube, (4) solvent feed tube, (5) fitting, (6) elbow, (7) turbine
assembly, (8) cap screw, (9) insulator, (10) shroud support,
(11) cleaning shroud, (12) shroud solvent drain, (13) nut,
(14) shroud solvent feed tube, (15) shroud solvent siphon.

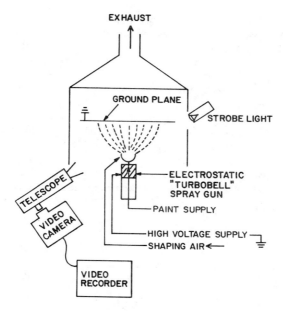

FIGURE 18.3   Arrangement for viewing the atomization zone and the spray
cloud within the paint spray booth.

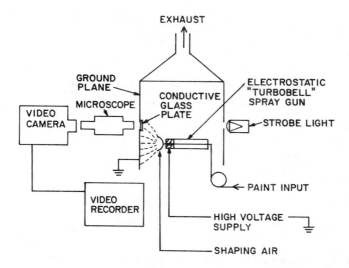

FIGURE 18.4   Arrangement for viewing the particle deposition and film
formation on a substrate.

flight could also be viewed with the high-speed video-microscopy optics.
In addition, the sizes of droplets collected on surfaces were measured.

The speed of the spray particles in flight was measured with laser-
Doppler instrumentation. The velocity distributions as well as the average
velocities were determined.

The charge carried by the spray was determined by electrical measurements of the charge-to-mass ratio, which, coupled with the particle
size data, gave information on the charge per particle.

The spray pattern geometry was observed on a large stationary target, with film thickness measurements across the target serving to characterize the pattern (Figure 18.5). Collected droplet size across the pattern
was determined by means of a shutter which permitted brief exposure of
collection surfaces (Figure 18.6). Appearance variations on panels were
determined from pattern-sampling arrays (Figure 18.6).

PATTERN OF STATIONARY TARGET

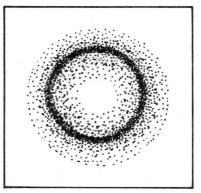

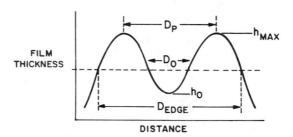

FIGURE 18.5   Schematic drawing of the donut-shaped spray pattern from the
rotary atomizer, and the variation of coating thickness across
the pattern.

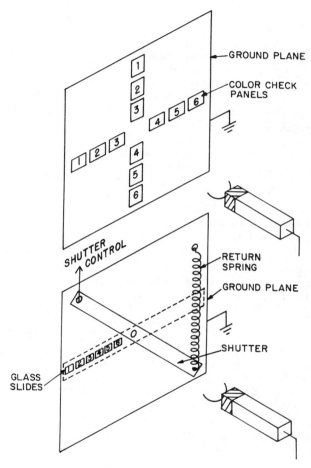

FIGURE 18.6   Schematic illustration of the arrangements used to sample different segments of the spray pattern. The numbers refer to the location of test panels.

The rheological, electrical and surface tension properties of the paints were determined by standard measurement methods.

MECHANICS OF ATOMIZATION

High-Speed Pictures

Profile views of the edge of the rotating bell during spraying were obtained using both high-speed video recording equipment for continuous motion analysis, and high-speed single-shot strobe photographs. Figure 18.7 shows the turbobell under several operating conditions with fluid filaments forming at the edge and subsequently breaking up into droplets.

The filament or ligament formation is characteristic of rotary atomizers even in the absence of electrostatic forces. The phenomenon is well documented in the spray drying literature [12,13]. The filaments form at high rotational speeds when the centrifugal force is large compared to the gravitational force and when the liquid feed rate is uniform and within a certain range. In the absence of high centrifugal forces, however, uniform filament formation will be induced by electrostatic forces, as shown by Miller [2] and Hines [3].

The filament length before breakup may range from three millimeters or less to more than ten millimeters, depending upon operating conditions and paint properties.

For the turbobell* used in the photographs of Figure 18.7, the inside edge had machined serrations with a spacing of 0.51 mm. The serrations influenced the filament spacing over a range of operating conditions. The darker filaments at 20,000 RPM shown in the two left photographs of Figure 18.7 have the serration spacing. A fine filament has formed between the primary filaments. At slow speeds, especially at high flow rates, the spacing became larger. The sheet formation in the lower right photograph contains ridges (dark bands) from the serration pattern.

At commercial operating speeds for the rotary atomizers of 20,000 to 30,000 RPM (with peripheral speeds of 75-115 m/s) the mechanical forces predominate over the electrostatic forces. Photographs and video micrographic studies at zero voltage showed little change in the filament formation from the high-voltage pattern.

Effect of Operating Conditions

Results of spray particle size measurements made at different operating conditions provided a quantitative evaluation of the effect of operating parameters. Figure 18.8 is a log-log graph of spray droplet size (mass mean

---

*Ransburg Part No. 20074-02.

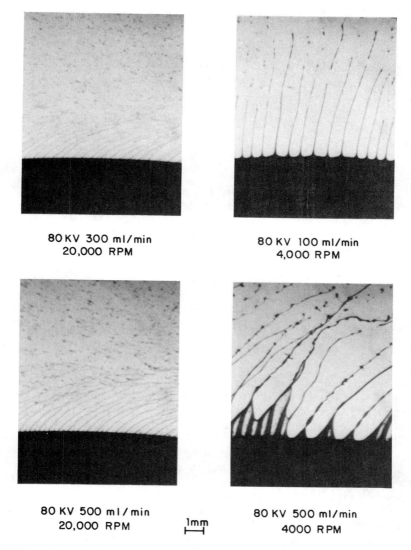

80 KV 300 ml/min
20,000 RPM

80 KV 100 ml/min
4,000 RPM

80 KV 500 ml/min
20,000 RPM

1mm
⊢——⊣

80 KV 500 ml/min
4000 RPM

FIGURE 18.7   High-speed flash photographs of the fluid filaments formed
at the edge of the rotary atomizer, under several conditions
of fluid flow rate and rotational speed.

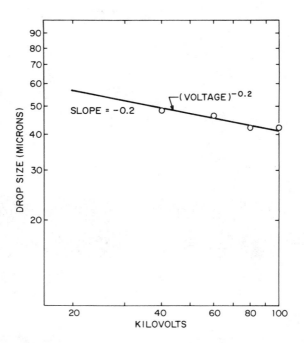

FIGURE 18.8   Effect of applied voltage on spray droplet size. The flow
rate was 250 ml/min and the atomizer speed was 20,000
RPM.

diameter) versus voltage. The results show that the particle size is inversely related to the voltage to the 0.2 power. This is a much smaller
electrostatic effect on atomization than reported by Hines [3] for the slow-
speed rotating-cone electrostatic atomizer, with a peripheral speed about
1/50th that of the high-speed bell.

The effect of turbine speed on spray droplet size is shown in Figure
18.9. Here the atomization has a relatively strong dependence on the turbine speed, being inversely related to the 0.7 power of the speed over the
measured range 7,000 to 24,000 RPM. Similarly, Figure 18.10 shows the
effect of droplet size on flow rate to be directly related to the 0.4 power of
the flow rate.

The power law exponents that we found for speed and feed rate are
within the range of similar exponents determined for rotary atomizers used
for spray drying [12]. The actual atomization process is sufficiently complex so that equations for the process are not known to have been derived
from basic principles. Aerodynamic forces from the peripheral speed and
centrifugal forces are interacting with the fluid mass, viscosity and surface
tension in a complex geometry.

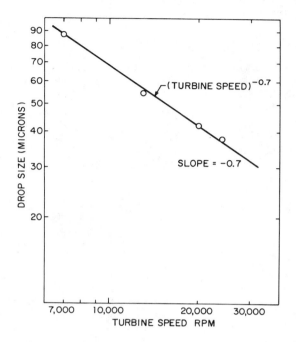

FIGURE 18.9   Effect of atomizer rotational speed on spray droplet size.
The flow rate was 250 ml/min and the voltage was 80 KV.

        The power law relations, however, provide a useful way to evaluate
quantitatively the effects of changes in spray parameters. Table 18.2
illustrates three cases where viscosity, turbine speed, and flow rate are
adjusted to give the same atomization. Particle size depends on viscosity
to the 0.2 power [12]. If viscosity is increased from 1.0 to 2.0 poise, an
increase in turbine speed from 20,000 RPM to 25,000 RPM would give the
same atomization, leaving the flow rate unchanged. Equivalently, a reduction in flow rate from 250 to 175 ml/min would be necessary to give the
same atomization if the turbine speed were unchanged.
        The magnitude of the spray particle size (mass mean diameter) for a
range of spray conditions for a particular paint is given in Table 18.3. As
shown by the data, the finest atomization is obtained at fast turbine speed,
low fluid flow rate, and high electrostatic voltage. Table 18.4 summarizes
the factors which lead to improved atomization [12].

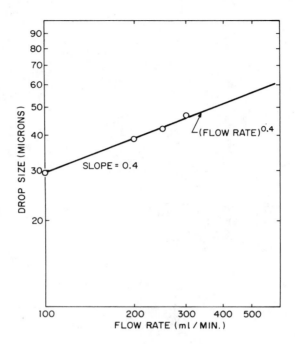

FIGURE 18.10   Effect of paint flow rate on spray droplet size. The
atomizer speed was 20,000 RPM and the voltage was
80 KV.

TABLE 18.2
Conditions for Equivalent Atomization

	I	II	III
Viscosity	1.0 poise	2.0	2.0
Turbine speed	20,000 RPM	25,000	20,000
Flow rate	250 ml/min	250	175

TABLE 18.3
Particle Size

Turbobell conditions		Particle size (μm)	
		80 KV	40 KV
Slow speed (7000 RPM)	Slow flow (100 ml/min)	65	75
Fast speed (20,000 RPM)	Slow flow	30	35
Slow speed	Fast flow (300 ml/min)	100	120
Fast speed	Fast flow	45	55

TABLE 18.4
Factors which Improve Atomization
($\rightarrow$ Smaller Droplet Size)

Increasing RPM
Decreasing flow rate

Decreasing viscosity
Decreasing surface tension
Decreasing density

Increasing voltage
Intermediate (?) resistivity

ELECTROSTATIC BREAKUP

While the dominant effect on atomization is from mechanical rather than electrical forces in the high-speed rotary atomizer, there is still a significant electrostatic contribution to spray particle size. In addition, the electrostatic field plays the vital role of imparting a charge on the spray particles which leads to the high collection efficiency of the electro-static spray process.

Some recent studies on the distintegration of liquid jets in an electrostatic field provide insight into the atomization process [11,14]. The jet length and the particle size were found to be a function of the electrical resistivity of the liquid. Mutoh et al. [14] describe four jet disintegration regimes (see Figure 18.11):

Regime A: Liquids with specific conductivities less than 10^{-11} ohm^{-1} cm^{-1} (resistance greater than 760 megohms with the Ransburg probe[*]) do not form droplets at all, but crawl up the outside wall of the capillary nozzle, clinging to it by electrostatic forces.

Regime B: Liquids with specific conductivities between 10^{-11} and 10^{-10} ohm^{-1} cm^{-1} (760 to 76 Ransburg megohms) form short and thick cylinders followed by disintegration into large droplets.

Regime C: Liquids with specific conductivities between 10^{-10} and 10^{-7} ohm^{-1} cm^{-1} (76 to .076 Ransburg megohms) form fine threads of liquid with fine droplets formed at a rapid rate by disintegration of the thread tip.

Regime D: Liquids with specific conductivities greater than 10^{-7} ohm^{-1} cm^{-1} (less than .076 Ransburg megohms) do not form a stable fine thread, as above. Instead, the whole thread breaks up into droplets when it becomes long enough.

These results have been schematically represented on an atomization scale versus resistivity in Figure 18.11. From the single jet studies [11, 14] we can therefore speculate that the electrostatic atomization from the rotary atomizer may be optimized in the region of 0.1 to 10 Ransburg megohms. Some initial data shown by the triangle points in Figure 18.11 appear to support this assumption.

The behavior of liquid jets in an electrostatic field appears to be related to the time constant for charge transfer [14], described by

$$\tau = \epsilon_o k_p$$

---

[*] Resistivity, ρ (ohm-m) = 1.31 R (Ransburg resistance, ohms).

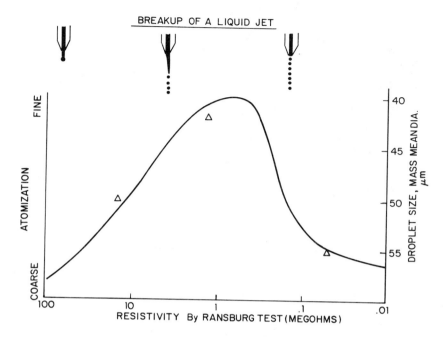

FIGURE 18.11   Effect of fluid resistivity on the disintegration of single
capillary jets induced by an electrostatic field. The triangle points represent results from the turbobell for
droplet size (as given by the scale at the right) versus
resistivity.

where

τ	= electrical time constant
ϵ_o	= permittivity of free space
k	= dielectric constant of material
ρ	= resistivity of material

Table 18.5 shows how the magnitude of the time constant is related to
the Ransburg resistance, for a dielectric constant of 2.4 (typical of hydrocarbons with weak polarity) and one of 32 (typical of liquids with high
polarity such as lower alcohols and ketones).

The time constant becomes important insofar as its magnitude relates to the time scale of dynamic events occurring during application.
Table 18.6 shows the magnitude of the time associated with paint application by high-speed rotary atomizers. We can calculate the time for a fluid
element to travel from the feed ring to the edge of the bell or disk [12, 13].
This time is approximately 500 μs, but the fluid also has additional time of

TABLE 18.5

Material Time Constant, Ranges: $\tau = \epsilon_o k_p$

R (megohms)	k = 2.4 $\tau(\mu s)$	k = 32 $\tau(\mu s)$
1000	28,000	370,000
100	2,800	37,000
10	280	3,700
1	28	370
.1	2.8	37
.01	.28	3.7

TABLE 18.6

Time Scale of Application Events

	Time (microsec.)	Resistance
Feed ring to edge of disk	500+	100 megohms or less
Edge of disk to breakup at end of filaments	100	10 – 0.1
Spray space between disk and substrate	50,000	---
Impact of spray droplets at substrate	5,000	100 or less

contact with the charged spray unit within the feed tube and feed ring. Thus more than 1000 μs may be the time available to charge the fluid surface by contact. A resistance of 100 Ransburg megohms or less would be expected to be adequate for charging.

At the edge of the atomizer, however, new surface is rapidly being formed. From the filament dimensions and the flow rates we know that a fluid element (particle) is within a filament for about 100 μs. This is the most rapid event of the application process. The paint resistance associated with this charging time is in the range of 0.1 to 10 Ransburg megohms, depending also on the dielectric constant.

We may therefore associate the high resistance atomization Regimes A and B with a fluid with insufficient conductivity to be electrostatically drawn into a filament. At an intermediate resistance as in Regime C, the filament formation time is comparable to the electrical charging time, and a field gradient can exist along the filament. This will elongate the filament more rapidly than without a field gradient and produce a finer thread of liquid before breakup [11,14], resulting in a finer particle size spray. When the resistance becomes lower, as in Regime D, the electrical charges move fast enough to reduce the field gradient along the filament, and the electrostatic stabilizing force vanishes. The actual disintegration process is also governed by the classical jet stability parameters of surface tension and viscosity [3,14].

The paint itself affects the electric field at the edge of the rotary atomizer and consequently the forces acting on the paint. This complicates the determination of the electrical effects [3]. Since the paint wets the edge of the atomizer and behaves like a conductor, the paint surface determines the electric field at the edge. The field is further modified by the space charge from the atomized droplets.

SPRAY PARTICLE CHARGING

Theoretical Considerations

The primary mechanism of spray particle charging is the breakup of the liquid which has been in contact with charged surfaces [3,6]. The amount of charge that can be carried by a liquid droplet is determined by the Rayleigh criterion [9,15]. This equates the maximum charge Q_{max} that a droplet of radius r and surface tension γ can sustain before the electrostatic surface force overcomes surface tension:

$$Q_{max} = 8\pi (\epsilon_o \gamma)^{1/2} r^{3/2}$$

where ϵ_o is the permittivity of free space. The Rayleigh limit has been experimentally verified over a wide range of droplet sizes and types of fluids [16,17].

A further limit on charge occurs when the surface electric field reaches such a high value that ion emission from the droplet can occur [8,9]. This requires high field strengths and small particle size. For liquid particles larger than about 0.2 μm field emission does not occur, and only the Rayleigh limit applies. Thus electrostatic paint sprays are dominated by the Rayleigh instability limit.

In addition to the "contact" charging from direct breakup of the charged liquid, the spray particles can acquire charge from corona-produced air ions in the spray space. The maximum charge which a droplet of radius r in an electric field E can attain in a corona ion field is [18]:

$$Q_{max} = 12\pi \ (k/k+2) \ \epsilon_o \ Er^2$$

where k = dielectric constant of droplet
      ϵ_o = permittivity of free space.

Field charging is related to the ordered motion of ions in the applied electric field which results in collisions between the ions and the more slowly moving particles. Another charging mechanism, "diffusion charging," results from ion-particle collisions due to the random thermal motion of the ions. Field charging is the dominant mechanism, however, for particles larger than about 1 μm [18].

The saturation charge from field charging will reach the Rayleigh limit for a conductive droplet when:

$$r_x = 0.444 \ \gamma / \epsilon_o \ E^2$$

If we assume a strong corona field of E = 600 KV/m and a surface tension of .030 N/m:

$$r_x = 4200 \ \mu m.$$

This indicates that particles as small as spray droplets cannot be charged to the Rayleigh stability limit by the ion field. (It is possible that the assumed field strength may be too low by an order of magnitude, reducing the r_x value to 40 μm or so. However, such a high field could only exist close to the sharp edge of the atomizer. The droplets leave that zone at such high speed that saturation charge cannot be attained.)

Charge/Mass Measurements

Measurements of charge per unit mass were made on several different paints over a range of operating conditions with the electrostatic turbobell. Samples were collected using a procedure similar to that described by Anestos et al. [6]. Particle size was also determined under similar conditions using a laser light-scattering technique.

Out present data are limited by the following two complications:

1. The distribution of charge is not uniform over the spray pattern. Reported results were based on a fixed geometry, but we know from additional measurements that the charge/mass tends to be higher near the center and lower at the outside of the spray cloud. The actual distribution may be a complicated function of the atomizer operating conditions and the paint properties.
2. The distribution of spray droplet size is not spatially uniform, either. As will be discussed later, smaller particles from the rotary atomizer tend to concentrate near the center of the spray cloud and larger particles to the outside. The magnitude of this effect is also dependent on atomizer operating conditions and paint properties.

In addition, the actual particle size distribution should be taken into account in determining the charge per particle, rather than simply the mass average size based on an assumed distribution.

The results of charge/mass measurements for four different voltages are shown in Table 18.7. The measured values are charge/mass and mass mean particle diameter. From these results we have calculated the charge per particle, the Rayleigh limiting charge per particle, and the number of electron charges per particle based on the measured charge/mass. It can be seen that the charging is much more effective at the highest voltage, but even at 100 KV the Rayleigh limit in this case is six times greater than the measured charge.

The charge/mass measurements versus voltage from Table 18.7 have been plotted in Figure 18.12. The results show a stronger dependence of charge/mass on voltage than the direct relationship (1.0 power) given by Hines [3] and Kelly [8]. The analysis by Hines of the charge transported by a fluid jet in an electric field yields the following equation (in SI units):

$$Q/M = 2E \ (\pi \sigma \epsilon_o^2 k^2 / \dot{M} d^2)^{1/3}$$

where Q/M = charge/mass ratio
$\quad\quad$ E = electric field
$\quad\quad$ σ = electrical conductivity of the fluid
$\quad\quad$ ϵ_o = permittivity of free space
$\quad\quad$ k = dielectric constant of the fluid
$\quad\quad$ $\dot{M}$ = mass flow rate
$\quad\quad$ d = density of the fluid

TABLE 18.7
Spray Particle Charge Vs. Voltage

Parameter	Applied electrostatic voltage, KV			
	100	80	60	40
Charge/mass (μC/g)	5.4	3.6	2.2	1.1
Droplet diameter (mass mean, μm)	42.2	42.1	46.4	48.3
Charge/particle (pC)	.211	.140	.114	.064
Rayleigh limit (pC)	1.26	1.26	1.45	1.54
Ratio: Raleigh/ measurement	6.0	9.0	12.7	24.1
Electron charges/ particle	1.3×10^6	$.88 \times 10^6$	$.71 \times 10^6$	$.40 \times 10^6$

It is reasonable to expect that the electric field E will not be directly proportional to the applied voltage. The geometry and properties of the fluid at the edge of the atomizer will modify the field. Our observations show a dependence of jet length on voltage with the unbroken jet length decreasing at the applied voltage increases. A similar observation has been reported by Miller [2]. The shortening of the jet also results in a proportional decrease in its resistance. As a result the current in the jets may increase more rapidly than the voltage. This is a possible explanation for the strong voltage dependence shown in Figure 18.12.

Table 18.8 shows the particle charging results obtained as a function of speed of the rotary atomizer. The charge/mass ratio remained relatively constant over a wide range of atomizer speeds, despite changes in the degree of atomization. The independence of charge/mass on particle size is a prediction of the analysis of Hines [3]. It results because the charge/mass ratio is proportional to the current flow to the atomizing jet divided by the mass flow rate, both of which are proportional to the square of the jet radius.

The lack of dependence of charge/mass on particle size substantiates the conclusion that the spray droplet charging is due to the breakup of the

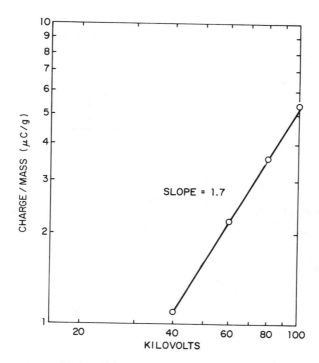

FIGURE 18.12   Effect of voltage on charge/mass ratio. The flow rate was
250 ml/min and the atomizer speed was 20,000 RPM.

charged liquid surface [3,6]. The charging mechanism is a conduction
process, resulting from the direct contact between the fluid and the charged
atomizer. If corona discharge played a significant role, the charging from
the corona ion field would show an inverse relationship between charge/
mass and droplet radius [6,18].

Table 18.9 illustrates the effect of flow rate on the charge/mass
measurements. A low flow rate leads to a higher charge/mass ratio as
predicted by the Hines equation. The experimental results appear to be
closer to the -0.66 power rather than the -0.33 power of the flow rate
given by that equation.

The effect of the resistivity of the paint on the charge/mass ratio is
shown in Figure 18.13. (The Ransburg resistance, R, is inversely proportional to the conductivity, through the relationship: Conductivity, ohm-1
m^{-1} = 0.76/R, ohm.) The data at two different voltages lead to a -0.2
power of R correlation with the charge/mass. The Hines equation predicts
a -0.33 power. No correction was made in our data, however, for possible
variations in the dielectric constant of the test paints.

TABLE 18.8
Spray Particle Charge Vs. Turbine Speed

Parameter	Turbine speed, RPM			
	24,000	20,000	13,000	7,000
Charge/mass (μC/g)	3.3	3.6	3.1	2.8
Droplet diameter (mass mean, μm)	37.9	42.1	54.9	87.6
Charge/particle (pC)	.093	.140	.27	.98
Ratio: Rayleigh/ measurement	11.5	9.0	7.0	3.9
Charge/mass field saturation (μC/g)	2.5	2.3	1.7	1.1

TABLE 18.9
Spray Particle Charge Vs. Fluid Flow Rate

Parameter	Flow rate (ml/min)	
	100	300
Charge/mass (μC/g)	6.8	2.9
Droplet diameter (mass mean, μm)	29.5	46.9
Charge/particle (pC)	.091	.156
Ratio: Rayleigh/ measurement	8.1	9.5
Charge/mass field saturation (μC/g)	3.2	2.0

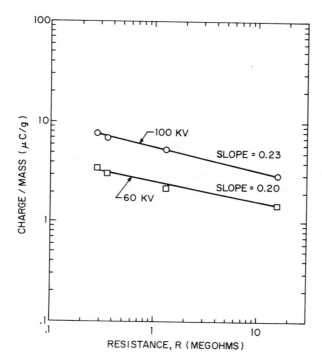

FIGURE 18.13   Effect of paint resistance on charge/mass ratio, for two
applied voltages. The flow rate was 250 ml/min and the
atomizer speed was 20,000 RPM.

SPRAY PARTICLE VELOCITY

        The spray particles from the rotary bell-shaped atomizer should
leave the edge at right angles to the axis of rotation due to the centrifugal
force. The speed at that point is defined by the peripheral speed, typically
in the range 75 to 120 m/s. The particles, initially in a turbulent zone,
rapidly decelerate as a result of air drag [12]. The particle trajectories
then curve toward the substrate as a result of the combined factors of
shaping air velocity, spray booth air velocity, and electrostatic forces.
Under the usual operating conditions, the electrostatic forces play the
dominant role in the particle trajectories, with the shaping air flow pro-
viding some added directionality, and with the booth ventilation flow being
a negligible factor.
        Shaping air velocities at the edge of the turbobell are shown in Figure
18.14, for a range of supply air pressures. A shaping air pressure of 20
psi (1.4 x 10^5 Pa) was used in much of our tests.

VELOCITY OF THE SHAPING AIR AT THE
EDGE OF THE BELL AT VARIOUS SUPPLY
PRESSURES

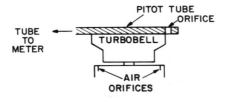

ALL READINGS WERE TAKEN WITH TURBINE AIR
OFF. AT ALL PRESSURE SETTINGS, ADJUSTED THE
PITOT TUBE TO GET THE HIGHEST STEADY READINGS

SUPPLY AIR PRESSURE PSIG	READINGS METERS / SEC
8	6.35
10	6.858
12	7.366
14	8.382
16	9.398
18	10.414
20	10.922
22	11.938
24	12.7

FIGURE 18.14   Shaping air velocities at the edge of the turbobell, for a
range of supply air pressures.

The air speeds and flow patterns within our laboratory spray booth
are shown in Figure 18.15 for 20 psi shaping air and in Figure 18.16 for
40 psi shaping air. The speeds are substantially slower than those resulting from conventional pneumatic spray guns, where air velocities approach
sonic speeds (300 m/s) near the gun tip and are about 5-15 m/s at application distances.

Spray droplet speeds were also measured directly within the spray
space using a laser-Doppler velocimeter. The velocity fluctuations were
quite large; for example, twice the standard deviation was in some cases
about the same magnitude as the average velocity. Under spray conditions
which produced a low charge/mass ratio, the particle speeds were found
to be about 1.5 m/s at a distance of 0.25 m from the atomizer. At 0.25 m,
the air speed measured with a pitot tube was approximately 1.3 m/s with

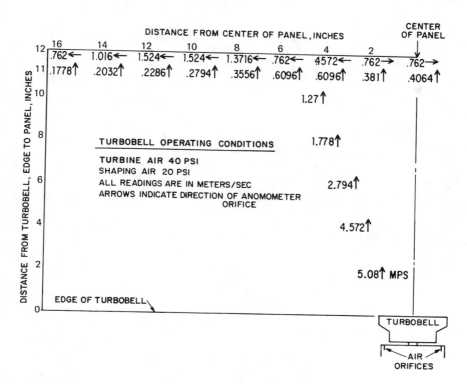

FIGURE 18.15 Air speeds and flow patterns within the laboratory spray booth. The shaping air pressure was 20 psi (1.4 x 10^5 Pa).

no spray present. The maximum particle speeds at this distance were about 3.5 m/s under high charge/mass conditions (low paint resistance, high voltage).

The air speed and particle speed measurements indicate that the particle travel time from the turbobell to the substrate is about 0.1 second. For a conventional pneumatic spray gun, similar measurements lead to a spray particle travel time of about 0.02 second under typical operating conditions. Thus the spray particles from the rotary electrostatic atomizer generally have a longer time of travel before reaching the substrate.

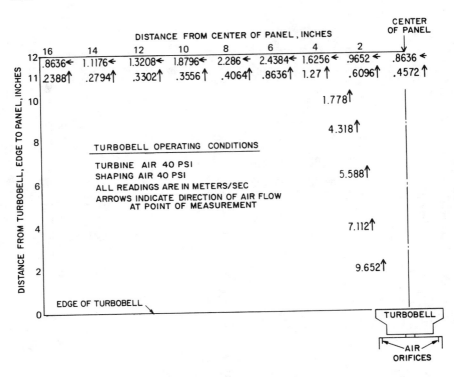

FIGURE 18.16   Air speeds and flow patterns within the laboratory spray
booth. The shaping air pressure was 40 psi (2.8 x 10⁵
Pa).

SPRAY PATTERN

As indicated by Figure 18.5, the spray pattern from the electrostatic
turbobell is donut-shaped on a stationary target. The dimensions of the
pattern are a function of all the variables that govern the spray particle
size, charge and velocity. The geometry of the electrostatic field, as well
as the mechanical factors introduced by the rotating atomizer and air flow
patterns, plays a role.

Interaction within the cloud of charged particles may also be significant. Electrostatic repulsion between droplets of like charge leads to
broadening of the spray cloud. Likewise the charge prevents coagulation
of approaching particles, maintaining the droplet dispersion. The details
of the interactions in charged particle streams have recently been described
in connection with ink-jet printing technology [19].

Figure 18.6 is a schematic illustration of the arrangements used to sample different segments of the stationary spray pattern. Panel arrays permitted assessment of the coating appearance across the pattern. Using the shutter, the deposited spray was sampled within controlled time intervals for the evaluation of particle size and charge/mass ratio.

The particle size of droplets collected on the substrate was found to be a function of the distance from the center of the pattern, as shown in Figure 18.17. The droplets near the center of the spray pattern were small, while larger particles were concentrated at the outside of the pattern. This behavior results from the centrifugal force and momentum keeping the larger particles in outside trajectories, while the competing force resulting from electrostatic attraction and shaping air flow more readily diverts the smaller particles toward the center of the pattern. Charge/mass measurements showed higher values nearer the center.

In general, the spray pattern became more uniform with some filling of the "donut hole" when the voltage was increased. Higher turbine speed for finer atomization and slower flow rates for both finer atomization and higher charging also gave more uniform patterns of deposition.

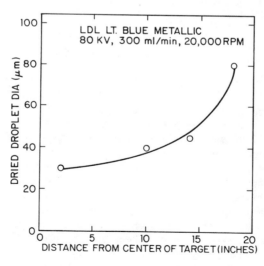

FIGURE 18.17   Particle size distribution across the spray pattern.

EFFICIENCY OF DEPOSITION

The efficiency with which charged particles are collected on a surface has been studied, especially with regard to electrostatic precipitation [18, 20] and the electrostatic powder coating process [4,5]. The theoretical basis for understanding the deposition process is well developed. In practice, however, the non-ideal geometries and large number of variables make it impractical to calculate the collection efficiency. The theory has utility, however, in guiding the direction that the operating parameters should take in order to gain improved efficiency.

The primary factors promoting collection efficiency are:

1. Large collecting surface area
2. Low gas stream velocity
3. High particle migration velocity

The particle migration velocity is governed by the electrical and inertial forces and the viscous resistance, with negligible effect from gravitational forces. The analysis as applied to electrostatic precipitation [18, 20] and to electrostatic powder coating [4] predicts greater efficiency for higher applied voltage and larger particle size when no significant inertial forces are present. For the case of the rotary atomizer, however, observations indicate that smaller particle size helps collection efficiency. This results from the initial particle inertial force acting at right angles to the coating direction.

METALLIC FLAKE EFFECTS

Orientation in Fluid Supply Line

The electrostatic spraying of metallic finishes leads to some special considerations. In the usual measurements of paint resistivity (such as with the Ransburg Paint Resistance Tester) the low-voltage test probe may not be significantly affected by dispersed metal particles. In the fluid line to the high-voltage atomizer, however, the metallic paint can develop increased conductivity due to flake orientation and settling.

The orientation develops from two sources: (1) preferred flake alignment from the fluid flow, and (2) orientation along the direction of the electric field. Increased flake concentration due to settling further increases the conductivity of the oriented system. The paint supply line then conducts an increasing current from the high-voltage end of the atomizer to the nearest ground, and may exceed the current limit of the power supply.

Atomization and Film Formation

High-speed video and photographic observation of the atomization zone at the edge of the turbobell showed no qualitative differences between metallic and non-metallic paints. The filament formation and disintegration retained the characteristics illustrated by Figure 18.7.

The collection of spray droplets at the ground plane, using the arrangement of Figure 18.6, showed the particle size variation across the spray pattern that was discussed before (refer to Figure 18.17) and an attendant variation in the concentration of metallic flake. The observations are illustrated in Figure 18.18. At 100 KV, small droplets predominated near the center of the pattern and the small droplets contained essentially no flake. The outside of the pattern consisted of large droplets within which were the flake particles. The spatial segregation of droplets and flake became less pronounced as the voltage was reduced.

Similar observations with a dispersion of mica flake gave the same results. This indicated that the electrical character of dispersed flake (metal or non-metal) was not a controlling factor. The flake versus droplet size interaction is probably a mechanical effect occurring at the point of filament breakup. As the fluid filament becomes increasingly thinner, breakup will occur whenever the size approaches the largest flake dimension. Smaller droplets will be formed when no flake comparable to the filament size happens to be present. Thus large flake will force a coarser atomization. Finer flake would be expected to be more uniformly dispersed over the range of particle sizes.

Figure 18.19 shows how fluid flow rate affected the droplet size and flake distribution across the spray pattern. At the lowest flow rate (50 ml/min) the droplet size was most uniform and small. At increasing flow rates, the particle size distribution became increasingly broad, with greater size differences between the inside and outside pattern. While difficult to see in the reproduction, the finer droplets did not contain flake.

SYMBOLS AND UNITS

d	mass density, kilograms/meters^3 (kg/m^3)
E	electric field, volts/meter (V/m)
k	dielectric constant (dimensionless)
M	mass, kilograms (kg)
$\dot{M}$	mass flow rate, kilograms/second (kg/s)
Q	charge, coulombs (C)
Q_{max}	maximum charge, coulombs (C)
R	resistance by Ransburg Paint Resistance Tester, ohms
RPM	rotational speed, revolutions/minute
r	radius, meters (m)

Bell and Hochberg

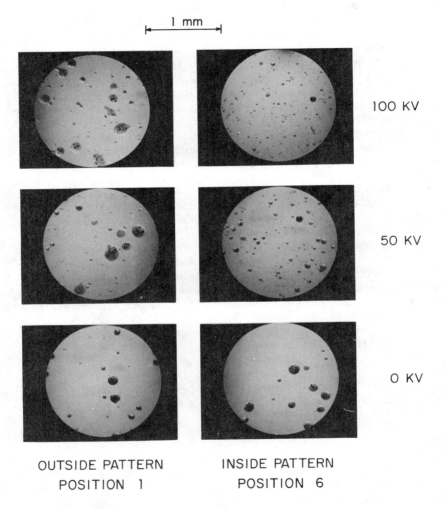

OUTSIDE PATTERN   INSIDE PATTERN
POSITION 1   POSITION 6

FIGURE 18.18   Effect of voltage on droplet size and distribution for a
               paint containing metallic flake. The droplets were collec-
               ted on glass slides as illustrated in the lower drawing of
               Figure 18.6. The flow rate was 250 ml/min and the
               atomizer speed was 20,000 RPM.

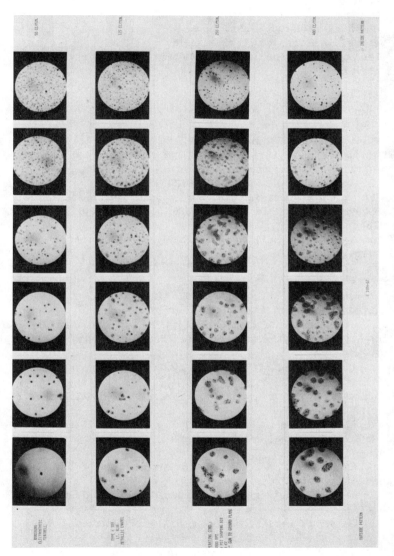

FIGURE 18.19   Effect of fluid flow rate on droplet size and flake distribution across the spray pattern. The rotational speed was 20,000 RPM and the voltage was 80 KV.

r_x droplet radius for field charge equal to Rayleigh limit, meters (m)

V   voltage, volts (V)

γ surface tension, newtons/meter (N/m)

ϵ_o permittivity of free space, coulombs/volt-meter (= 8. 85 $\times 10^{-12}$ C/V-m)

ρ electrical resistivity, ohm-meter

σ electrical conductivity, ohm^{-1} $meter^{-1}$

τ time constant, seconds (s); time to acquire 63% of equilibrium charge

REFERENCES

1.   S. Suslik, Industrial Finishing, 57(1), 20 (1981).
2.   E. P. Miller, in A. D. Moore (Ed.), Electrostatics and Its Applications, Wiley-Interscience, New York, 1973, Chap. 11, p. 250.
3.   R. L. Hines, J. Appl. Phys. , 37, 2730 (1966).
4.   S. Wu, Polym.-Plast. Technol. Eng. , 7, 119 (1976).
5.   A. Golovoy, J. Paint Technol. , 45(580), 42(1973); 45(585), 68, 74 (1973).
6.   T. C. Anestos, J. E. Sickles, and R. M. Tepper, IEEE Trans. Industry Applic. , 1A-13, 168 (1977).
7.   R. M. Tepper, J. E. Sickles, and T. C. Anestos, IEEE Trans. Industry Applic. , 1A-13, 177 (1977).
8.   A. J. Kelly, J. Appl. Phys. , 49, 2621 (1978).
9.   A. J. Kelly, J. Appl. Phys. , 47, 5264 (1976).
10. C. D. Hendricks, in A. D. Moore (Ed.), Electrostatics and Its Applications, Wiley-Interscience, New York, 1973, Chaps. 2-4, p. 8.
11. D. W. Horning and C. D. Hendricks, J. Appl. Phys. , 50, 2614 (1979).
12. K. Masters, Spray Drying Handbook (3rd ed.), Godwin, London, 1979.
13. W. R. Marshall, Jr. , Chem. Eng. Progr. , 50, 31 (1954).
14. M. Mutoh, S. Kaieda, and K. Kamimura, J. Appl. Phys. , 50, 3174 (1979).
15. L. D. Landau and E. M. Lifshitz, Electrodynamics of Continuous Media, Pergamon Press, London, 1960, p. 35.
16. A. Doyle, D. R. Moffett, and B. Vonnegut, J. Colloid Sci. , 19, 136 (1964).
17. J. W. Schweizer and D. N. Hanson, J. Colloid Interface Sci. , 35, 417 (1971).
18. S. Oglesby, Jr. , and G. B. Nichols, in A. C. Stern (Ed.), Air Pollution (3rd ed.), Vol. IV, Academic Press, New York, 1977, p. 189.

19.   L. Kuhn and R. A. Myers, Scientific American, $\underline{240}$, 162 (April 1978).

20.   S. Oglesby, Jr. and G. B. Nichols, Electrostatic Precipitation, Dekker, New York, 1978.

TOPICAL PROBLEMS IN THE FIELD OF WATER
SOLUBLE RESINS AND EMULSIONS

Rolf Zimmermann

Hoechst AG
Frankfurt am Main, West Germany

ELECTRO POWDER COATING

The procedure of electro powder coating first developed at Shinto-Paint of Japan [1] is a new and interesting variation of electrodeposition. A ready to use EPC material basically consists of an aqueous binder solution pigmented with a formulated epoxide powder coating with a cross-linking agent. Cathodic deposition then follows the mechanisms known from the standard deposition process. This method combines the favorable properties of powder coatings and of the electrodeposition procedure.

In such a formulation the aqueous binder acts as a wetting and dispersing aid for the powder. In addition it functions as a carrier for deposition on negatively charged metal substrates. Its basic groups after neutralization with organic acids promote water solubility. Under baking conditions cross-linking takes place between the binder and the powder and hardener.

Best results are obtained with epoxide powders with particle sizes in the range of 5-10 μm, by which sedimentation is prevented and clear polymer films are possible. Epoxide powders or epoxide/polyester combinations are preferred.

Currently, blocked aromatic or aliphatic diisocyanates are used as hardener.

Fields of Application

Basically, the EPC process covers the same areas as standard electro coatings. The throwing power, however, is inferior. Therefore, only flat objects can be coated completely. Consequently, the system should only be used where the throwing power is of minor importance. Currently available coatings are suitable for one coat systems applied on metal parts which require a high standard of chemical and corrosion resistance.

The kinetics of film formation of EPC coatings differs significantly from standard electro coatings. Films from anionic or cathodic paints are limited to 15-35 μm in thickness, depending on the paint formulation. They are deposited in 2-4 min. The throwing power is good and film thickness rises with increasing voltage and increasing bath temperature. EPC paint films of 40-70 μm are obtained within much shorter deposition periods of 10-60 sec.

The difference in throwing power and film thickness between standard electrodeposition materials and electro powder coatings is caused by the difference in specific wet film resistance and by the dependence of this film resistance on deposition conditions [2].

The remarkable porosity of wet EPC films is brought about by the hydrogen generated during deposition which needs to penetrate through the coating layer. Approximately 0.2 m^3 per cm^2 of film are formed, which means thirty times the film volume. There is every indication to believe that an increase in film resistance with increasing bath temperature and voltage is found when the number of pores is decreasing at lower film viscosities.

The attainable film thickness of 40-70 μm which could previously only be achieved with powder coatings opens up new applications for the electrodeposition process.

Reverse Process

Due to the limited throwing power, the film thickness level, the film properties, and particularly the high stone chipping resistance, EPC qualifies for the reverse coating process. By this process the exterior surface is first electropowder-coated, and after pre-drying the interior is covered by a standard cationic paint. Then complete curing of EPC cross-linking of the cathodic material are carried out in a joing baking process. Currently, reverse coating is applied by two Japanese automotive companies and at two pilot plants in Germany.

Processing under Laboratory Conditions

The binder is reduced with deionized water to a solid content of 10-16%. In order to obtain good initial wetting the resin solution is slowly added to the powder while the remaining resin can be added more rapidly.

Tank Characteristics:

Solids	10-16%
pH	5.0-5.5
Specific conductivity	800-1200 μS cm^{-1}
Bath temperature	25°C
Deposition voltage	Approx. 200 volt
Coating time	20 sec.

As a pretreatment for the iron substrate a proper zinc phosphatisation (Bonder 130) is most suitable. Film thickness depends on the voltage, deposition time, bath temperature, metal pretreatment and the binder to powder ratio, as shown in Figures 19.1 through 19.6. Table 19.1 summarizes the main film properties obtained under the conditions mentioned.

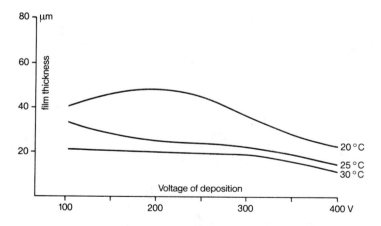

FIGURE 19.1   Influence of bath temperature and voltage of deposition on film thickness. Deposition time 20 sec. Ratio of powder to binder 1:1.

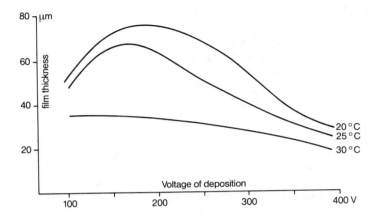

FIGURE 19.2   Influence of bath temperature and voltage of deposition on
film thickness. Deposition time 20 sec. Ratio of powder
to binder 1.4:1.

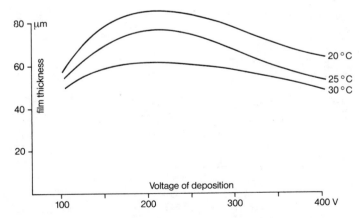

FIGURE 19.3   Influence of bath temperature and voltage of deposition on
film thickness. Deposition time 20 sec. Ratio of powder
to binder 1.8:1.

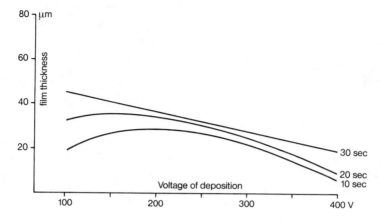

FIGURE 19.4   Influence of voltage and time of deposition on film thickness.
Bath temperature 25°C.   Ratio of powder to binder 1.7:1.

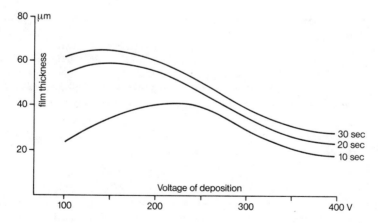

FIGURE 19.5   Influence of voltage and time of deposition on film thickness.
Bath temperature 25°C.   Ratio of powder to binder 1.4:1.

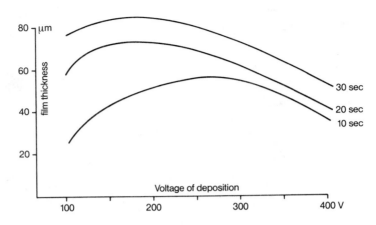

FIGURE 19.6   Influence of voltage and time of deposition on film thickness.
Bath temperature 25°C.   Ratio of powder to binder 1.8:1.

Practical Use of Reverse Coating

The reverse process used in practice may be summarized as follows:
metal cleaning and pretreatment, EPC, pre-drying, cathodic ED, and
baking.

The zinc phosphated car bodies are passed through the EPC tank and
are predried in an oven for 10 min. at approximately 80°C. Afterwards
uncovered areas are coated in an electrophoretic tank containing a cationic
binder. Hardening of the film is achieved by a two step stoving procedure:
10 min. at 80°C (dryoff), 30 min. at 180°C (bake), 5 min. cooling zone.
In order to obtain good levelling and to avoid blistering, monitoring of the
time/temperature program is important. This is explained in Figure 19.7.

Conclusion

EPC is a new coating process for metal surfaces. Using the ED
technology thick films of excellent corrosion resistance are obtained. Due
to low throwing power the procedure is limited to readily accessible surfaces. EPC, however, is most suitable for the reverse coating process.
Most advantageous are high paint utilization and low energy consumption
together with other favorable properties known from standard ED.

TABLE 19.1
Paint Film Properties

Test	Degreased steel sheet	Zinc bonder Bonder 130
Impact test	50 inch-pound	Over 80 inch-pound
Erichsen test	9 mm	9 mm
Mandrel test	O. K.	O. K.
Lattice cut	1.0	1.0
Salt spray test ASTM B 117 750 hours	Cross cut rusting 8 mm	Cross cut rusting 1 mm
Stone chipping resistance salt spray test 360 hours	O. K.	O. K.
Water resistance test 50°C	Over 8 weeks	Over 8 weeks
Humidity resistance test 100% RH, 50°C	Over 1000 hours	Over 1000 hours
20% Sodium hydroxide room temperature	Over 8 weeks	Over 8 weeks
20% Sulfurice acid room temperature	Over 8 weeks	Over 8 weeks
Xylene room temperature	Over 8 weeks	Over 8 weeks

During our first pilot-tests, we experienced some problems such as cratering, unsatisfactory corrosion protection at sharp edges or in the areas of overlapping, and uneven film surfaces making sanding necessary. Meanwhile a modification of the binder and the cross-linking agent has brought some improvements, at least on a laboratory scale, and we are optimistic for the future.

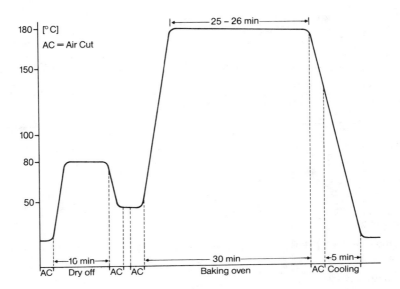

FIGURE 19.7   Stoving time/temperature program.

ANAPHORETIC RESINS OF IMPROVED THROWING POWER

In recent years the American and Japanese automobile industry made
a relatively rapid change from anaphoresis to cationic coatings. Meanwhile, European automotive producers partially went the same way, however, with anaphoretic resins still in use, especially resins based on
polybutadiene/maleic anhydride adducts. Such coating compositions show
a balanced combination of properties especially as far as throwing power,
corrosion resistance and stone chipping resistance are concerned.

In comparison with anaphoresis cathodic electrodeposition has the
following advantages:

1.   Improved throwing power
2.   Improved salt spray resistance at low film thickness
3.   Improved alkali stability
4.   Excellent edge coverage in coating shaped metal parts

Nevertheless, the following disadvantages should also be mentioned:

1.   The film properties depend heavily on the pretreatment of
     the metal.

2. Optimum corrosion protection is only achieved by the use of lead pigments.
3. Larger amounts of by-products are found at baking.
4. Minor resistance to stone chipping in a three or four layer set up is found in particular when the stoving temperature for the primer is not selected properly.

Some research has been done looking in particular for an improvement of the throwing power of anaphoretic systems [3]. As a first step addition of maleic anhydride to polybutadiene with a large amount of "cis" double bonds is carried out. In the presence of an alcohol a ring opening reaction of the anhydride follows forming a half ester. The free carboxyl groups then are neutralized with an organic amine, to give

$$
\cdots-\overset{H_2}{C}\diagdown\overset{H_2\ H}{C-C}\diagdown\overset{trans-}{\underset{H/}{C-C}}\overset{H_2\ H_2}{C-C}\diagdown\overset{H_2}{C}-\cdots
$$

$$
\underset{H\ \ H}{C=C}\qquad\underset{trans-\ H}{C-C}\qquad\underset{H_2H\ \ H}{C=C}
$$

$$
\underset{H}{C-C}\overset{H_2}{}
$$

$$
O=\overset{}{C}\qquad\overset{}{C}-O
$$

$$
\overset{|}{OR}\qquad\overset{|}{O}^{(-)}
$$

$$
[HN(R)_3]^{(+)}
$$

For cross-linking a resole type carboxylic acid obtained from the reaction of bisphenol-A with chloroacetic acid is preferred. The properties of these carboxylic acids can be varied by the introduction of partially etherified methylol groups. The structure of such a cross-linking agent [4] is

$$
ROH_2C-\bigcirc(HO)-\overset{CH_3}{\underset{CH_3}{C}}-\bigcirc-O-CH_2-C\overset{O}{\underset{OH}{\diagup}}
$$

with ROH_2C groups

R=H or Alkyl

Priority was given to the improvement of throwing power and corrosion resistance of thin films. Throwing power is influenced by the chemical structure of the polymer binder, and the pigmentation. The parameters determining the throwing power of the resin are the electrical resistance of the freshly deposited coating, and the electrical conductivity of the aqueous binder.

The electrical resistance is governed by the glass transition temperature of the polymer, the average molecular weight, the gaseous electrolysis products, and the remaining amount of polar and therefore electrically conductive groups.

Electrical conductivity of the aqueous binder solution is determined by the acid number of the resin and its neutralization rate, the size of the cation, the dissociation constant of the carboxyl groups, the temperature and the concentration in the dipping tank. Taking these parameters into consideration it has recently been possible to develop anaphoretic resins with a throwing power similar to cationic binders. The anti-corrosive properties required at low film thickness, however, could not be achieved.

Table 19.2 compares the film properties of standard and newer anaphoretic resins with cathodic systems as a function of bath characteristics.

TABLE 19.2
Comparison of Different Electrophoretic Resins

	Anaphoretic resins (basis: polybutadiene)		Kataphoretic resins (modified epoxide resins)
	Old	New	
Acid value	120	92	--
pH value	7.0	6.8	6.5
Specific bath conductivity (μS cm^{-1})	3850	1970	1400
Voltage of deposition (V)	170	300	290
Throwing power (cm)	12.5	20.0	20.0
Stone chipping resistance	Good	Good	Satisfying
Edge covering	Satisfying	Satisfying	Very good
Salt spray resistance in hr (film thickness 20 μm)			
Iron-steel sheet	ca. 200	ca. 200	400
Iron-passivated steel sht.	ca. 500	ca. 400	500
Zinc-phosphated steel sheet	500	500	500

The anaphoretic resin with improved throwing power is already being used by several European automobile producers and they will in our opinion retain their importance in various areas of automotive production and general industrial application.

AIR-DRYING WATER THINNABLE ALKYD RESINS

Water-thinnable systems are mainly polymers with carboxylic or basic nitrogen groups. Water solubility is achieved by salt formation with organic amines or acids. Another possibility to synthesize water soluble resins is to introduce strongly hydrophilic groups, such as polyethylene glycol ethers or their combination with carboxylic groups into the polymer chain.

1. Polyanion resin

2. Polycation resin

3. Resin with hydrophilic groups

For easy processing of water-borne paints based on these types of resins, the presence of relatively large amounts of organic solvents is necessary to overcome the abnormal viscosity in the course of the dilution with water.

We therefore put our efforts into the development of alkyd dispersions and colloidal systems. We felt that these products would need only small quantities of organic co-solvents and amines for improved water reducibility. In other words we expected a normal rheological behavior of these systems when being thinned with water.

Dispersions of Air-Drying Alkyd Resins

Basically, alkyd resins can be made emulsifiable in water by two methods: (1) use of external emulsifiers, and (2) formulating resins which are emulsifiable themselves.

External emulsifiers show two disadvantages; they usually show less chemical resistance than the polymer, and due to low molecular weights

they are easily extracted from the apint film. These disadvantages are
significantly reduced by incorporating the emulsifying groups into a polymer. Polyethylene glycol or polypropylene glycol are useful diols for this
purpose, but they plasticize the paint film and in addition increase its
water sensitivity. We therefore employed modified polyethylene glycols
according to the following formula

The polyethylene glycol is etherified with a phenolic resole. Remaining methylol groups are reacted with tung oil. This emulsifying compound
then is used for alkyd resin condensation. To produce an emulsion, this
emulsifying resin is combined with an alkyd without polyether groups. The
resulting dispersions have solid contents of 42%. They contain 0.7% of
amines and as a levelling agent 3.5% of butyl glycol.

Dispersions of this type are used in air-drying and baking primers,
and also in colored one-layer coatings. The paint films show excellent
water resistance and corrosion protection. The properties of the dispersions, and above all their hydrolytic stability, are improved by reacting
the emulsifying agent not with an alkyd, but with a carboxyl functional
acrylic resin instead. The hydrophilic carboxylic groups are separated
by extended CH_2-chains. Thus saponification of ester linkages is reduced.

Air-Drying Colloidal Systems

   Proceeding from alkyd emulsions to colloidal systems containing particles much smaller in size, the amount of emulsifier had to be increased. Unfavorable influences of the emulsifiers on film resistance were avoided using emulsifiers which are crosslinked with the polymer during the drying process.

   Products of this nature are for instance polycondensates containing conjugated fatty acid groups. These resins are blended with air-drying alkyds insoluble in water [5]. The composition is determined by the end use.

   Since emulsifier and emulsifiable resins must be compatible to yield clear films, the range of products which can be combined is limited. To get homogeneous coatings, the oil length of both resin components which is responsible for their drying characteristics, must be adapted to each other.

   The colloidal systems are prepared in the following way. First, a clear solution is formed by mixing emulsifier and emulsifiable resin components while adding water miscible organic solvents and neutralizing agents such as amines. Prior to processing this solution is thinned with water producing colloidal systems in which the emulsifier dissolves in water emulsifying the water insoluble resin.

   The basic structure of these colloidal air-drying systems is shown below.

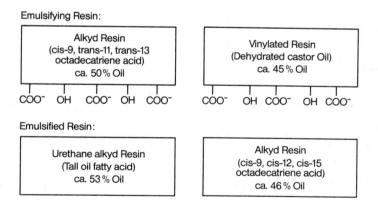

Emulsifying Resin:

| Alkyd Resin (cis-9, trans-11, trans-13 octadecatriene acid) ca. 50% Oil |
| Vinylated Resin (Dehydrated castor Oil) ca. 45% Oil |

COO⁻ OH COO⁻ OH COO⁻   COO⁻ OH COO⁻ OH COO⁻

Emulsified Resin:

| Urethane alkyd Resin (Tall oil fatty acid) ca. 53% Oil |
| Alkyd Resin (cis-9, cis-12, cis-15 octadecatriene acid) ca. 46% Oil |

   Table 19.3 summarizes the performance of the products representing the present state of development.

   We have been engaged in the development of fast drying water soluble alkyds for high gloss paints. Up to now we have not succeeded in producing water soluble vehicles for house paints which are comparable in drying characteristics and film-build to solvent-based long-oil alkyds.

TABLE 19.3
Comparison of the Properties of Water–Thinnable
Air Drying Systems

Properties	Alkyd dispersions	Colloidal solutions	
		Resin I	Resin II
Application	Priming (metal parts of all kinds)	Primer (farm machinery, radiators)	Priming and top coats (wood and metal)
Coating procedure	Spraying	Dipping and spraying	Brushing and spraying
Special properties	High corrosion protection	Rapidly drying	Suitable for top coats
Pigmentability	2	1	1
Processing	3	1–2	1
Drying	2–3	1	2–3
Gloss	1–2	2–3	1–2
Yellowing	4	2	1–2
Recoatability	1	1	1
Water resistance	1	1	1
Storage stability 20°C	1	1–2	2
Corrosion protection (salt spray test)	1	1	1
Solvent content of coating	4%	14%	16%
Amine content	0.7%	1.3%	1.7%

1 = Very good
5 = Unsatisfactory

Better results can be obtained with gloss paints based on acrylic dispersions with structure and wet-adhesion [6].

Water Based Epoxide/Phenolic Resin Dispersions

For many years organic solutions of resole type phenolics combined with polyvinyl butyrals and/or long chain epoxide resins have been used as binders in interior and exterior coatings for food containers. The films formed after curing provide high flexibility, excellent metal adhesion, and chemical resistance.

Chemical resistance was also found with other types of protective coatings when the resin blends mentioned were filled with indifferent pigments like titanium dioxide, iron oxide or barytes. Applications such as tank linings, interior coatings of chemical apparatus or interior pipe coatings, are quite common. The performance is ruled by the phenolic resin/plasticizing component ratio.

Systems of this nature are usually applied by roller coating or spraying. Ethyl glycol, glycol ethers, esters, diacetone alcohol, other ketones, or aromatics are employed as solvents or thinners.

To minimize environmental pollution and to avoid costly afterburners during application, coating materials with reduced solvent content or completely water based became more and more important.

Due to this demand we strongly engaged in the development of water based epoxide/phenolic resin dispersions [7]. This work finally resulted in emulsions which generally meet the specifications required.

The water based dispersions discussed consist of blends of amine modified epoxides with a tailored phenolic resole. Water solubility is achieved by the addition of acids which means we are dealing with a cationic dispersion. (See structure on page 374.)

Special efforts had to be made for developing the phenolic component. By special processing methods we succeeded in synthesizing resoles which under baking conditions do not split off by-products. They also generate unusual storage stability at elevated temperatures and still meet the cross-linking requirements for standard systems. Curing can be completed in 5-20 min. at 170°-200° C. Since the ingredients used in this resin may be varied within wide limits a great variety in application properties is feasible.

The properties obtained on an interior can coating in comparison with a standard solvent based system are demonstrated in Table 19.4.

In general water based dispersions show flow problems and adhesive failures when they are applied by roller coating. By using grooved rolls, reverse coating methods or spray application those difficulties can be eliminated. It is obvious, however, that further developmental work is necessary to improve application performance.

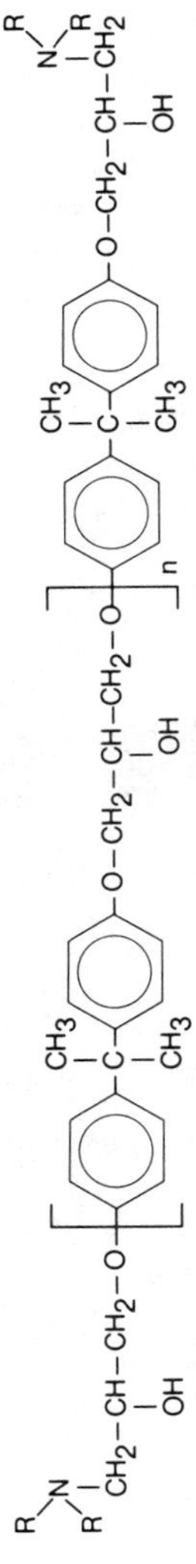

Modified epoxide resin

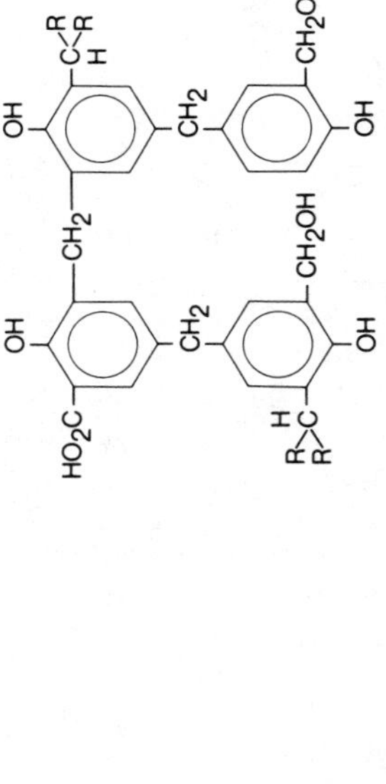

Phenolic resin